VHDL-2008

The Morgan Kaufmann Series in Systems on Silicon
Series Editor: Wayne Wolf, Georgia Institute of Technology

The Designer's Guide to VHDL, Second Edition
Peter J. Ashenden

The System Designer's Guide to VHDL-AMS
Peter J. Ashenden, Gregory D. Peterson, and Darrell A. Teegarden

Modeling Embedded Systems and SoCs
Axel Jantsch

ASIC and FPGA Verification: A Guide to Component Modeling
Richard Munden

Multiprocessor Systems-on-Chips
Edited by Ahmed Amine Jerraya and Wayne Wolf

Functional Verification
Bruce Wile, John Goss, and Wolfgang Roesner

Customizable and Configurable Embedded Processors
Edited by Paolo Ienne and Rainer Leupers

Networks-on-Chips: Technology and Tools
Edited by Giovanni De Micheli and Luca Benini

VLSI Test Principles & Architectures
Edited by Laung-Terng Wang, Cheng-Wen Wu, and Xiaoqing Wen

Designing SoCs with Configured Processors
Steve Leibson

ESL Design and Verification
Grant Martin, Andrew Piziali, and Brian Bailey

Aspect-Oriented Programming with the e Verification Language
David Robinson

Reconfigurable Computing: The Theory and Practice of FPGA-Based Computation
Edited by Scott Hauck and André DeHon

System-on-Chip Test Architectures
Edited by Laung-Terng Wang, Charles E. Stroud, and Nur A. Touba

Verification Techniques for System-Level Design
Masahiro Fujita, Indradeep Ghosh, and Mukul Prasad

VHDL-2008: Just the New Stuff
Peter J. Ashenden and Jim Lewis

VHDL-2008

Just the New Stuff

Peter J. Ashenden
Consultant
Ashenden Designs

Jim Lewis
Director of Training
SynthWorks Design, Inc.

ELSEVIER

AMSTERDAM • BOSTON • HEIDELBERG • LONDON
NEW YORK • OXFORD • PARIS • SAN DIEGO
SAN FRANCISCO • SINGAPORE • SYDNEY • TOKYO

Morgan Kaufmann Publishers is an imprint of Elsevier

MORGAN KAUFMANN PUBLISHERS

Publishing Director	Joanne Tracy
Publisher	Denise E.M. Penrose
Senior Acquisitions Editor	Charles Glaser
Publishing Services Manager	George Morrison
Senior Production Editor	Dawnmarie Simpson
Assistant Editor	Matthew Cater
Production Assistant	Lianne Hong
Cover Designer	Dennis Schaefer
Cover Image	Scott Tysick/Masterfile
Composition	Peter J. Ashenden
Copyeditor	JC Publishing
Proofreader	Janet Cocker
Indexer	Joan Green
Interior printer	Sheridan Books, Inc.
Cover printer	Phoenix Color, Inc.

Morgan Kaufmann Publishers is an imprint of Elsevier.
30 Corporate Drive, Suite 400, Burlington, MA 01803, USA

∞ This book is printed on acid-free paper.

Library of Congress Cataloging-in-Publication Data
Ashenden, Peter J.
    VHDL-2008 : just the new stuff / Peter J. Ashenden, Jim Lewis.
        p. cm.
    Includes index.
    ISBN 978-0-12-374249-0 (pbk. : alk. paper) 1. VHDL (Computer hardware description language)
I. Lewis, Jim. II. Title.

TK7885.7.A846 2007
621.39'2--dc22
                                        2007039499

ISBN: 978-0-12-374249-0

For information on all Morgan Kaufmann publications,
visit our Web site at *www.mkp.com* or *www.books.elsevier.com*

Printed and bound in the United Kingdom

Transferred to Digital Print 2011

Contents

Preface

VHDL is defined by IEEE Standard 1076, *IEEE Standard VHDL Language Reference Manual* (the VHDL LRM). The original standard was approved in 1987. IEEE procedures require that standards be periodically reviewed and either reaffirmed or revised. The VHDL standard was revised in 1993, 2000, and 2002. In each revision, new language features were added and some existing features enhanced. The aim in each revision was to improve the language as a tool for design and verification of digital systems.

Since the 2002 revision, there have two parallel efforts to further develop the language. The first was the VHDL Procedural Interface (VHPI) Task Force, a subcommittee of the IEEE P1076 Working Group. The VHPI Task Force prepared an interim amendment to the standard, formally approved by IEEE in March 2007. The amendment is titled *IEEE 1076c, Standard VHDL Language Reference Manual—Amendment 1: Procedural Language Application Interface.*

In the second effort, during 2004 and 2005, the P1076 Working Group undertook preliminary work toward a new revision of the standard. In June 2005, the board of Accellera approved formation of a Technical Committee (TC) to continue that work, funded jointly by Accellera and TC members directly. The Accellera VHDL-TC worked intensively between September 2005 and June 2006, producing a new draft of the LRM, P1076/D3.0. This draft was a full revision of the VHDL standard, defining numerous new and enhanced language features, incorporating minor clarifications and corrections, and including the VHPI specification from IEEE 1076c. The language defined by this draft is informally called VHDL-2006. The draft was published for trial use by implementers and users during the period from June 2006 to June 2007. Feedback has been rolled into a subsequent draft to be forward to the P1076 Working Group for IEEE standardization. The final version will be informally called VHDL-2008.

The aim of this book is to introduce the new and changed features of VHDL-2008 in a way that is more accessible to users than the formal definition in the LRM. We describe the features, illustrate them with examples, and show how they improve the language as a tool for design and verification. We assume you are already familiar with earlier versions of VHDL, specifically VHDL-2002 and VHDL-93. These versions are described comprehensively in *The Designer's Guide to VHDL, Second Edition*, by Peter Ashenden, also published by Morgan Kaufmann Publishers. We hope that the present book will be helpful not only to early adopters of the new language version, but also to tool implementers seeking to understand what it is they have to implement.

In addition to the information presented in this book, additional reference information is available at the authors' web sites:

- www.ashenden.com.au
- www.SynthWorks.com

Acknowledgments

We sincerely thank David Bishop of Kodak, Bill Logan of Rockwell Collins, and Chuck Swart of Mentor Graphics for their technical review of various chapters of this book. Their comments led to significant improvement in our explanations and correction of coding errors. Presenting code examples for language features yet to be implemented in tools is a risky business. Having "human compilers" check the code is most valuable. Any remaining errors are, of course, ours.

We would also like to thank Chuck Glaser, our editor at Elsevier, for his encouragement to develop this book. Chuck has a keen sense of what the market needs, and we are happy to take his advice.

Finally, we would like to thank you, the reader, in advance for any comments and corrections. We would love to hear from you, by email at vhdl-book@ashenden.com.au. We will maintain a list of errata on the web sites mentioned above.

Chapter 1

Enhanced Generics

We start our tour of the new features in VHDL-2008 with one of the most significant changes in the language, enhanced generics. All earlier versions of VHDL since VHDL-87 have included generic constants, which are interface constants for design entities and components. They are widely used in models to represent timing parameters and to control the widths of vector ports. When we instantiate an entity or component, we supply values for the generic constants for that instance. The generic constants in the generic list are called the *formal generics*, and the values we supply in the generic map are called the *actual generics*. Most of the time, generic constants are referred to just as "generics," since the only kind of generics are constants.

In VHDL-2008, generics are enhanced in several significant ways. First, we can declare generic types, subprograms, and packages, as well as generic constants. Second, we can declare generics on packages and subprograms, as well as on entities and components. The rationale for extending generics in these ways is to increase productivity by allowing us to declare reusable entities, packages, and subprograms that deal with different types of data and that can be specialized to perform different actions. In this chapter, we will describe each of the new kinds of generics and the new places in which we can declare generics.

1.1 Generic Types

Generic types allow us to define a type that can be used for ports and internal declarations of an entity, but without specifying the particular type. When we instantiate the entity, we specify the actual type to be used for that instance. As we will see later, generic types can also be specified for packages and subprograms, not just for entities and components.

We can declare a *formal generic type* in a generic list in the following way:

type *identifier*

The identifier is the name of the formal generic type, and can be used within the rest of the entity in the same way as a normally declared type. When we instantiate the entity, we specify a subtype as the *actual generic type*. This can take the form of a type name, a type name followed by a constraint, or a subtype attribute.

EXAMPLE 1.1   *A generic multiplexer*

A multiplexer selects between two data inputs and copies the value of the selected input to the output. The behavior of the multiplexer is independent of the type of data on the inputs and output. So we can use a formal generic type to represent the type of the data. The entity declaration is:

```
entity generic_mux2 is
  generic ( type data_type );
  port    ( sel : in bit; a, b : in data_type;
            z : out data_type );
end entity generic_mux2;
```

The name data_type is the formal generic type that stands for some type, as yet unspecified, used for the data inputs a and b and for the data output z. An architecture body for the multiplexer is:

```
architecture rtl of mux2 is
begin
  z <= a when sel = '0' else b;
end architecture rtl;
```

The assignment statement simply copies the value of either a or b to the output z. It is sensitive to all of the inputs. So whenever a, b, or sel change, the assignment will be re-evaluated. In any instance of the multiplexer, changes on a and b are determined using the predefined equality operator for the actual type in that instance.

We can instantiate the entity to get a multiplexer for bit signals as follows:

```
signal sel_bit, a_bit, b_bit, z_bit : bit;
...

bit_mux : entity work.generic_mux2(rtl)
  generic map ( data_type => bit )
  port map    ( sel => sel_bit, a => a_bit, b => b_bit,
                z => z_bit );
```

Similarly, we can instantiate the same entity to get a multiplexer for signals of other types, including user-defined types.

```
type msg_packet is record
  src, dst : unsigned(7 downto 0);
  pkt_type : bit_vector(2 downto 0);
  length   : unsigned(4 downto 0);
  payload  : byte_vector(0 to 31);
  checksum : unsigned(7 downto 0);
end record msg_packet;
signal pkt_sel : bit;
```

```
signal pkt_in1, pkt_in2, pkt_out : msg_pkt;
...

pkt_mux : entity work.generic_mux2(rtl)
  generic map ( data_type => msg_packet )
  port map    ( sel => pkt_sel,
                a => pkt_in1, b => pkt_in2, z => pkt_out );
```

VHDL-2008 defines a number of rules covering formal generic types and the ways they can be used. The formal generic type name can potentially represent any constrained type, except a file type or a protected type. The entity can only assume that operations available for all such types are applicable, namely: assignment; allocation using **new**; type qualification and type conversion; and equality and inequality operations. The formal generic type cannot be used as the type of a file element or an attribute. Moreover, it can only be used as the type of an explicitly declared constant or a signal (including a port) if the actual type is not an access type and does not contain a subelement of an access type. For signals, the predefined equality operator of the actual type is used for driver update and event detection.

If we have a formal generic type T, we can use it to declare signals, variables, and constants within the entity and architecture, and we can write signal and variable assignments for objects of the type. For example, the following shows signals declared using T:

```
signal s1, s2 : T;
...

s1 <= s2 after 10 ns;
```

and the following shows variables declared using T:

```
variable v1, v2, temp : T;
...

temp := v1;  v1 := v2;  v2 := temp;
```

Since signal and variable declarations require constrained subtypes, the actual type provided in an instance must be a constrained type if the formal type is used in this way. If the actual type is not constrained, an error occurs in the instantiation. If the formal generic type is not used in any way requiring it to be constrained, then the actual type in an instance need not be constrained.

For both variables and signals, the default initial value is determined using the actual type in an instance, using the normal rules for the actual type. Thus, if the actual type is a scalar type, the default initial value is the leftmost value of the type, and if the actual type is a composite type, the default initial value is an aggregate of the default initial values for the respective element types.

Declaring constants of a formal generic type might at first seem impossible, since we can't specify an initial value if we don't know the actual type. However, we can use the formal generic type to declare a formal generic constant, and then use that within the entity, for example:

```
entity e is
  generic ( type T; constant init_val : T );
  port    ( ... );
end entity e;

architecture a of e is
begin
  p : process is
    variable v : T := init_val;
  begin
    ...
  end process p;
end architecture a;
```

The actual value for the generic constant is provided when the entity is instantiated, and must be of the type specified as the actual generic type. For example, we might instantiate the entity e within a larger design as follows:

```
my_e : entity work.e(a)
  generic map ( T => std_ulogic_vector(3 downto 0),
                init_val => "ZZZZ" );
```

We can also use this technique to provide values for initializing variables and signals declared to be of the formal generic type. Note that the generic list in this entity makes use of one generic (T) in the declaration of another generic (init_val). This was illegal in previous versions of VHDL, but is now legal in VHDL-2008 (see Section 9.1).

One thing that we cannot do with formal generic types is apply operations that are not defined for all types. For example, we cannot use the "+" operator to add to values of a formal generic type, since the actual type in an instance may not be a numeric type. Similarly, we cannot perform array indexing, or apply most attributes. This may at first seem an onerous restriction, but it does mean that a VHDL analyzer can check the entity and architecture for correctness in isolation, independently of any particular instantiation. It also means we don't get any surprises when we subsequently analyze an instance of the entity. Fortunately, as we will see in Section 1.5, there are ways of providing operations to an instance for use on values of the actual type.

EXAMPLE 1.2   *Illegal use of formal generic types*

Suppose we want to define a generic counter that can be used to count values of types such as **integer**, **unsigned**, **signed**, and so on. We can declare the entity as follows:

```
entity generic_counter is
  generic ( type      count_type;
            constant reset_value : count_type );
```

```
  port     ( clk, reset : in  bit;
              data         : out count_type );
end entity generic_counter;
```

We might then try to define an architecture as:

```
architecture rtl of generic_counter is
begin
  count : process (clk) is
  begin
    if rising_edge(clk) then
      if reset = '1' then
        data <= reset_value;
      else
        data <= data + 1;   -- Illegal
      end if;
    end if;
  end process count;
end architecture rtl;
```

The problem is that the "+" operator to add 1 to a value is not defined for all types that might be supplied as actual types. Hence, the analyzer will indicate an error in the expression where the operator is applied. To illustrate why this should be an error, suppose some time after the entity and architecture have been written, we try to instantiate them in a design as follows:

```
type traffic_light_color is (red, yellow, green);
...


cycle_lights : entity work.generic_counter(rtl)
  generic map ( count_type  => traffic_light_color,
                reset_value => red )
  port map    ( ... );
```

The process in the instance would have to apply the "+" operator to a value of the actual generic type, in this case, **traffic_light_color**. That application would fail, since there is no such operator defined. We will revise this example in Section 1.5 to show how to supply such an operator to the instance.

Note in passing that the process in this example reads the value of the **out**-mode parameter **data** in an expression. While this was illegal in earlier versions of VHDL, it is legal in VHDL-2008 (see Section 6.3).

When we declare a generic constant in a generic list, we can specify a default value that is used if no actual value is provided in an instance. For generic types, there is no means of specifying a default type. That means that we must always specify an actual type in an instance. Since the type of objects in VHDL is considered to be a very important property, the language designers decided to insist on the actual type being explicitly specified.

1.2 Generic Lists in Packages

One of the new places in which we can write generic lists in VHDL-2008 is in package declarations. A package with a generic list takes the form:

```
package identifier is
  generic ( ... );

  ...  -- declarations within the package

end package identifier;
```

The package body corresponding to such a package is unchanged; we don't repeat the generic list there. Within the generic list, we can declare formal generic constants and formal generic types, just as we can in a generic list of an entity or component. We can then use those formal generics in the declarations within the package.

A package with a generic list is called an *uninstantiated package*. Unlike a simple package with no generic list, we cannot refer to the declarations in an uninstantiated package with selected names or use clauses. Instead, the uninstantiated package serves as a form of template that we must instantiate separately. We make an instance with a *package instantiation* of the form:

```
package identifier is new uninstantiated_package_name
  generic map ( ... );
```

The identifier is the name for the package instance, and the generic map supplies actual generics for the formal generics defined by the uninstantiated package. If all of the formal generics have defaults, we can omit the generic map to imply use of the defaults. (As we mentioned in Section 1.1, if any of the formal generics is a generic type, it cannot have a default. In that case, we could not omit the generic map in the package instance.) Once we have instantiated the package, we can then refer to names declared within it with selected names and use clauses with the instance name as the prefix.

For now, we will assume that the uninstantiated package and the package instance are declared as design units and stored in a design library. We will refine this assumption in Section 1.3.

EXAMPLE 1.3 *A package for stacks of data*

We can write a package that defines a data type and operations for fixed-sized stacks of data. A given stack has a specified capacity and stores data of a specified type. The capacity and type are specified as formal generics of the package, as follows:

```
package generic_stacks is
  generic ( size : positive; type element_type );
```

```
type stack_array is array (0 to size-1) of element_type;
type stack_type is record
  SP    : integer range 0 to size-1;
  store : stack_array;
end record stack_type;

procedure push (s : inout stack_type; e : in  element_type);
procedure pop  (s : inout stack_type; e : out element_type);

end package generic_stacks;
```

The corresponding package body is:

```
package body generic_stacks is

  procedure push (s : inout stack_type; e : in  element_type) is
  begin
    s.store(s.SP) := e;
    s.SP := (s.SP + 1) mod size;
  end procedure push;

  procedure pop  (s : inout stack_type; e : out element_type) is
  begin
    s.SP := (s.SP - 1) mod size;
    e := s.store(s.SP);
  end procedure pop;

end package body generic_stacks;
```

The uninstantiated package defines types **stack_array** and **stack_type** for representing stacks, and operations to **push** and **pop** elements. The formal generic constant **size** is used to determine the size of the array for storing elements, and the formal generic type element_type is the type of elements to be stored, pushed and popped.

We cannot refer to items in this uninstantiated package directly, since there is no specification of the actual size and element type. Thus, for example, we cannot write the following:

```
use work.generic_stacks.all;  -- Illegal
...
variable my_stack : work.generic_stacks.stack_type;  -- Illegal
```

Instead, we must instantiate the package and provide actual generics for that instance. For example, we might declare the following as a design unit for a CPU design:

```
library IEEE; use IEEE.numeric_std.all;
package address_stacks is new work.generic_stacks
  generic map ( size => 8,
                element_type => unsigned(23 downto 0) );
```

If we analyze this instantiation into our working library, we can refer to it in other design units, for example:

```
architecture behavior of CPU is
  use work.address_stacks.all;

  . . .
begin
  interpret_instructions : process is
    variable return_address_stack : stack_type;
    variable PC : unsigned(23 downto 0);

    . . .
  begin

    . . .
    case opcode is
      when jsb => push(return_address_stack, PC);
                  PC <= jump_target;
      when ret => pop(return_address_stack, PC);

      . . .
    end case;

    . . .
  end process interpret_instructions;
end architecture behavior;
```

This architecture includes a use clause that makes names declared in the package instance **address_stacks** visible. References to **stack_type**, **push** and **pop** in the architecture thus refer to the declarations in the **address_stacks** package instance.

We can declare multiple instances of a given uninstantiated package, each with different actual generics. The packages instances are distinct, even though they declare similarly named items internally. For example, we might declare two instances of the **generic_stacks** package from Example 1.3 as follows:

```
package address_stacks is new work.generic_stacks
  generic map ( size => 8,
                element_type => unsigned(23 downto 0) );

package operand_stacks is new work.generic_stacks
  generic map ( size => 16, element_type => real );
```

If we then wrote a use clause in a design unit:

```
use work.address_stacks.all, work.operand_stacks.all;
```

the names from the two package instances would all be ambiguous. This is an application of the existing rule in VHDL that, if two packages declare the same name and both are "used," we cannot refer to the simple name, since it is ambiguous. Instead, we need to use selected names to distinguish between the versions declared in the two package instances. So, for example, we could write:

```
use work.address_stacks, work.operand_stacks;
```

to make the package names visible without prefixing them with the library name **work**, and then declare variables and use operations as follows:

```
variable return_address_stack : address_stacks.stack;
variable PC                    : unsigned(23 downto 0);
variable FP_operand_stack      : operand_stacks.stack;
variable TOS_operand           : real;
...
address_stacks.push(return_address_stack, PC);
operand_stacks.pop(FP_operand_stack, TOS_operand);
```

An important aspect of VHDL's strong-typing philosophy is that two types introduced by two separate type declarations are considered to be distinct, even if they are structurally the same. Thus the two types declared as

```
type T1 is array (1 to 10) of integer;
type T2 is array (1 to 10) of integer;
```

are distinct, and we cannot assign a value of type T1 to an object of type T2. This same principle applies to formal generic types. Within an entity or a package that declares a formal generic type, that type is considered to be distinct from every other type, including other formal generic types. So, for example, we cannot assign a value declared to be of one formal generic type to an object declared to be of another formal generic type.

The fact that two formal generic types are distinct can lead to interesting situations when the actual types provided are the same (or are subtypes of the same base type). Ambiguity can arise between overloaded operations declared using the formal generic types. This kind of situation is not likely to happen in common use cases, but it is worth exploring to demonstrate the way overloading works in the presence of formal generic types.

Suppose we declare a package with two formal generic types, as follows:

```
package generic_pkg is
  generic ( type T1; type T2 );

  procedure proc ( x : T1 );
  procedure proc ( x : T2 );
  procedure proc ( x : bit );

end package generic_pkg;
```

Within the package, T1 and T2 are distinct from each other and from the type **bit**, so the procedure **proc** is overloaded three times. The uninstantiated package can be analyzed without error. If we instantiate the package as follows:

```
package integer_boolean_pkg is new work.generic_pkg
  generic map ( T1 => integer, T2 => boolean );
```

we can successfully resolve the overloading for the following three calls to procedures in the package instance:

```
work.integer_boolean_pkg.proc(3);
work.integer_boolean_pkg.proc(false);
work.integer_boolean_pkg.proc('1');
```

On the other hand, if we instantiate the package as

```
package integer_bit_pkg is new work.generic_pkg
  generic map ( T1 => integer, T2 => bit );
```

the following call is ambiguous:

```
work.integer_bit_pkg.proc('1');
```

It could be a call to the second or third of the three overloaded versions of **proc** in the package instance. Similarly, if we instantiate the package as

```
package integer_integer_pkg is new work.generic_pkg
  generic map ( T1 => integer, T2 => integer );
```

the following call is ambiguous:

```
work.integer_integer_pkg.proc(3);
```

This could be a call to the first or second of the three overloaded versions of **proc**. The point to gain from these examples is that overload resolution depends on the actual types denoted by the formal generic types in the instances. Depending on the actual types, calls to overloaded subprograms may be resolvable for some instances and ambiguous for others.

The final aspect of packages with generic lists is that we can also include a generic map in a package, following the generic list. Such a package is called a *generic-mapped package*, and has the form

```
package identifier is
  generic ( ... );
  generic map ( ... );

  ... -- declarations within the package

end package identifier;
```

The generic list defines the generics, and the generic map aspect provides actual values and type for those generics. While VHDL-2008 allows us to write a generic-mapped package explicitly, we would not normally do so. Rather, the feature is included in the language as a definitional aid. An instantiation of an uninstantiated package is defined in terms of an equivalent generic-mapped package that is a copy of the uninstantiated package, together with the generic map from the instantiation. This is analogous to the way in which an entity instantiation is defined in terms of a block statement that merges the generic and port lists of the entity with the generic map and port map of the instantiation. Since generic-mapped packages are not a feature intended for regular use, we won't dwell on them further. We simply mention them here to raise awareness, since the occasional error message from an analyzer might hint at them.

1.3   Local Packages

In earlier versions of VHDL, packages can only be declared as design units. They are separately analyzed into a design library, and can be referenced by any other design unit that names the library. Thus, they are globally visible. In VHDL-2008, packages can also be declared locally within the declarative region of an entity, architecture, block, process, subprogram, protected type body, or enclosing package. This allows the visibility of the package to be contained to just the enclosing declarative region. Moreover, since declarations written in a package body are not visible outside the package, we can use local packages to provide controlled access to locally declared items.

EXAMPLE 1.4   *Sequential item numbering*

Suppose we need to generate test cases in a design, with each test case having a unique identification number. We can declare a package locally within a stimulus-generator process. The package encapsulates a variable that tracks the next identification number to be assigned, and provides an operation to yield the next number. The process outline is:

```
stim_gen : process is

  package ID_manager is
    impure function get_ID return natural;
  end package ID_manager;

  package body ID_manager is
    variable next_ID : natural := 0;
    impure function get_ID return natural is
      variable result : natural;
    begin
      result := next_ID;
      next_ID := next_ID + 1;
      return result;
    end function get_ID;
```

```
    end package body ID_manager;

    ...

begin

    ...
    test_case.ID := ID_manager.get_ID;
    ID_manager.next_ID := 0;   -- Illegal

    ...
end process stim_gen;
```

The variable **next_ID** is declared in the package body, and so is not visible outside the package. The only way to access it is using the **get_ID** function provided by the package declaration. This is shown in the first assignment statement within the process body. The package name is used as a prefix in the selected name for the function. The second assignment statement is illegal, since the variable is not visible at that point. The package provides a measure of safety against inadvertent corruption of the data state.

We can write use clauses for locally declared packages. Thus, we could follow the package declaration in this example with the use clause

```
use ID_manager.all;
```

and then rewrite the assignment in the process as

```
test_case.ID := get_ID;
```

By writing the package locally within the process, it is only available in the process. Thus, we have achieved greater separation of concerns than had we written the package as a design unit, making it globally visible. Moreover, since the package is local to a process, there can be no concurrent access by multiple processes. Thus, the encapsulated variable can be an ordinary non-shared variable. If the package were declared as a global design unit, there could be concurrent calls to the **get_ID** function. As a consequence, the variable would have to be declared as a shared variable of a protected type. This would significantly complicate the design.

_____

As Example 1.4 illustrates, if a package declared within a declarative region requires a body, then the body must come after the package declaration in the same region. If the enclosing region is itself a package, then we write the inner package declaration within the enclosing package declaration, and the inner package body within the outer package body. If the inner package requires a body, then the outer package requires a body as a consequence.

A locally declared package need not be just a simple package. It can be an uninstantiated package with a generic list (or, indeed, a generic-mapped package with both generic list and generic map). In that case, we must instantiate the package so that we can refer to items in the instance. The same rules apply to locally declared uninstantiated packages and instances as apply to globally declared packages.

EXAMPLE 1.5 *Package for wrapping items with item numbers*

We can revise the package from Example 1.4 to make it deal with test cases of
generic type, and to wrap each test case in a record together with a unique ID number. The numbers are unique across test cases of all types. We achieve this by keeping the previous package as an outer package encapsulating the **next_ID** variable.
Within that package, we declare an uninstantiated package for wrapping test cases.
The process outline containing the packages is:

```
stim_gen : process is

  package ID_manager is

    package ID_wrappers is
      generic ( type test_case_type );
      type wrapped_test_case is record
        test_case : test_case_type;
        ID        : natural;
      end record wrapped_test_case;
      impure function wrap_test_case
                      ( test_case : test_case_type )
                      return wrapped_test_case;
    end package ID_wrappers;

  end package ID_manager;

  package body ID_manager is

    variable next_ID : natural := 0;

    package body ID_wrappers is
      impure function wrap_test_case
                      ( test_case : test_case_type )
                      return wrapped_test_case is
        variable result : wrapped_test_case;
      begin
        result.test_case := test_case;
        result.ID := next_ID;
        next_ID := next_ID + 1;
        return result;
      end function wrap_test_case;
    end package body ID_wrappers;

  end package body ID_manager;

  use ID_manager.ID_wrappers;
```

```
    package word_wrappers is new ID_wrappers
      generic map ( test_case_type => unsigned(32 downto 0) );
    package real_wrappers is new ID_wrappers
      generic map ( test_case_type => real );

    variable next_word_test : word_wrappers.wrapped_test_case;
    variable next_real_test : real_wrappers.wrapped_test_case;

begin
    . . .
    next_word_test := word_wrappers.wrap_test_case(X"0440CF00");
    next_real_test := real_wrappers.wrap_test_case(3.14159);

    . . .
end process stim_gen;
```

The process declares two instances of the uninstantiated package ID_wrappers, one for a test-case type of **unsigned**, and another for a test-case type of **real**. The process then refers to the **wrapped_test_case** type and the **wrap_test_case** function declared in each instance.

Example 1.5 exposes a number of important points about packages. First, a package declared within an enclosing region is just another declared item, and is subject to the normal scope and visibility rules. In the example, the ID_wrappers package is declared within an enclosing package, and so can be referred to with a selected name and made visible by a use clause.

Second, in the case of package instantiations, any name referenced within the uninstantiated package keeps its meaning in each instance. In the example, the name **next_ID** referenced within the uninstantiated package ID_wrappers, refers to the variable declared in the ID_manager package. So, within each of the package instances, **word_wrappers** and **real_wrappers**, the same variable is referenced. Importantly, had the process also declared an item called **next_ID** outside the packages but before the instances, that name would not be "captured" by the instances. They still refer to the same variable nested within the ID_manager package. The only exception to this rule for interpreting names is that the name of the uninstantiated package itself, when referenced within the package, is interpreted in an instance as a reference to the instance. This allows us to use an expanded name for an item declared within the uninstantiated package, and to have it interpreted appropriately in the instance. The rules for name interpretation illustrate quite definitely that package instantiation is different in semantics from file inclusion, as is used for C header files. The benefit of the VHDL-2008 approach is that names always retain the meaning they are given at the point of declaration, and so we avoid unwanted surprises.

The third point is that local instantiation of an uninstantiated package is a common use case, whether the uninstantiated package be locally declared, as in the example, or globally declared as a design unit. The advantage of local instantiation is that it allows use of a locally declared type as the actual for a formal generic type. Were local instantiation not possible, the actual type would have to be declared in a global package in order to use it in a global package instantiation. Thus, local instantiation improves modularity and information hiding in a design.

EXAMPLE 1.6 *Local stack package instantiation*

In Example 1.3, we declared an uninstantiated package for stacks as a design unit. We can instantiate the package to deal with stacks of a type declared locally within a subprogram that performs a depth-first search of a directed acyclic graph (DAG) consisting of vertices and edges, as follows:

```
subprogram analyze_network ( network : network_type ) is

  type vertex_type is ...;
  type edge_type   is ...;
  constant max_diameter : positive := 30;

  package vertex_stacks is new work.generic_stacks
    generic map ( size => max_diameter,
                  element_type => vertex_type );
  use vertext_stacks.all;

  variable current_vertex   : vertex_type;
  variable pending_vertices : stack_type;

begin
  ...
  push(pending_stacks, current_vertex);
  ...
end subprogram analyze_network;
```

The data types used to represent the DAG for analyzing a network are the local concern of the subprogram. By instantiating the **generic_stacks** package locally, there is no need to expose the data types outside the subprogram.

1.4 Generic Lists in Subprograms

The second new place in which we can write generic lists in VHDL-2008 is in subprogram (procedure and function) declarations. A procedure with a generic list takes the form:

```
procedure identifier
  generic   ( ... )
  parameter ( ... ) is
  ... -- declarations
begin
```

```
    ...  -- statements
end procedure identifier;
```

Similarly, a function with a generic list takes the form:

```
function identifier
  generic  ( ... )
  parameter ( ... ) return result_type is
  ...  -- declarations
begin
  ...  -- statements
end function identifier;
```

We use terminology analogous to that for packages to refer to subprograms with generics. Thus, a subprogram with a generic list is called an *uninstantiated subprogram*. Note that the new keyword **parameter** is included to make the demarcation between the generic list and the parameter list clear. For backward compatibility, including the keyword is optional. We expect that designers will omit it for subprograms without generics and include it or not as a matter of taste for uninstantiated subprograms.

VHDL allows us to declare a subprogram in two parts, one consisting just of the specification, and the other consisting of the specification together with the body. We can separate a subprogram in this way within a given declarative part, for example, in order to declare mutually recursive subprograms. In the case of subprograms declared in packages, we are required to separate the subprogram specification into the package declaration and to repeat the specification together with the subprogram body in the package body. In the case of uninstantiated subprograms, the generic list is part of the subprogram specification. Thus, if we separate the declaration, we must write the generic list and parameter list in the specification, and then repeat both together with the body. Using a text editor to copy and paste the specification into the body makes this easy.

We cannot call an uninstantiated subprogram directly. We can think of it as a template that we must instantiate with a *subprogram instantiation* to get a real subprogram that we can call. For a procedure, the instantiation is of the form:

```
procedure identifier is new uninstantiated_procedure_name
  generic map ( ... );
```

and for a function, the instantiation is of the form

```
function identifier is new uninstantiated_function_name
  generic map ( ... );
```

In both cases, the identifier is the name for the subprogram instance, and the generic map supplies actual generics for the formal generics defined by the uninstantiated subprogram. If all of the formal generics have defaults, we can omit the generic map to imply use of the defaults. Once we have instantiated the subprogram, we can then use the instance name to call the instance.

EXAMPLE 1.7   *Generic swap procedure*

The way in which we swap the values of two variables does not depend on the types of the variables. Hence, we can write a swap procedure with the type as a formal generic, as follows:

```
procedure swap
  generic   ( type T )
  parameter ( a, b : inout T ) is
  variable temp : T;
begin
  temp := a; a := b; b := temp;
end procedure swap;
```

We can now instantiate the procedure to get versions for various types:

```
procedure int_swap is new swap
  generic map ( T => integer );
procedure vec_swap is new swap
  generic map ( T => bit_vector(0 to 7) );
```

and call them to swap values of variables:

```
variable a_int, b_int : integer;
variable a_vec, b_vec : bit_vector(0 to 7);
...

int_swap(a_int, b_int);
vec_swap(a_vec, b_vec);
```

We can't just call the **swap** procedure directly, as follows:

```
swap(a_int, b_int); -- Illegal
```

since it is an uninstantiated procedure. Note also that we can't instantiate the **swap** procedure with an unconstrained type as the actual generic type, since the procedure internally uses the type to declare a variable. Thus, the following would produce an error:

```
procedure string_swap is new swap generic map ( T => string );
```

since there is no specification of the index bounds for the variable temp declared within swap.

EXAMPLE 1.8 *Setup timing check procedure*

Suppose we are developing a package of generic operations for timing checks on signals. We include a generic procedure that determines whether a signal meets a setup time constraint. The package declaration is:

```
package timing_pkg is
  procedure check_setup
    generic ( type signal_type;
              type clk_type; clk_active_value : clk_type;
              T_su : delay_length )
    ( signal s : signal_type; signal clk : clk_type );
  ...
end package timing_pkg;
```

The package body contains a body for the procedure:

```
package body timing_pkg is
  procedure check_setup
    generic ( type signal_type;
              type clk_type; clk_active_value : clk_type;
              T_su : delay_length )
    ( signal s : signal_type; signal clk : clk_type ) is
  begin
    if clk'event and clk = clk_active_value then
      assert s'last_event >= T_su
        report "Setup time violation" severity error;
    end if;
  end procedure check_setup;
  ...
end package body timing_pkg;
```

We can now instantiate the procedure to get versions that check the constraint for signals of different types and for different setup time parameters:

```
use work.timing_pkg.all;
procedure check_normal_setup is new check_setup
  generic map ( signal_type => std_ulogic,
                clk_type => std_ulogic,
                clk_active_value => '1',
                T_su => 200ps );
procedure check_normal_setup is new check_setup
  generic map ( signal_type => std_ulogic_vector,
                clk_type => std_ulogic,
                clk_active_value => '1',
                T_su => 200ps );
procedure check_long_setup is new check_setup
  generic map ( signal_type => std_ulogic_vector,
```

```
                      clk_type => std_ulogic,
                      clk_active_value => '1',
                      T_su => 300ps );
```

Note that the procedure **check_normal_setup** is now overloaded, once for a
std_ulogic parameter and once for a **std_ulogic_vector** parameter. We can apply
these functions to signals of **std_ulogic** and **std_ulogic_vector** types, as follows:

```
signal status : std_ulogic;
signal data_in, result : std_ulogic_vector(23 downto 0);
. . .

check_normal_setup(status, clk);
check_normal_setup(result, clk);
check_long_setup(data_in, clk);
. . .
```

In each case, the active value for the clock signal and the setup time interval
value are bound into the definition of the procedure instance. We do not need to
provide the values as separate parameters.

VHDL-2008 allows us to declare uninstantiated subprograms and to instantiate them
in most places where we can currently declare simple subprograms. That includes declaring uninstantiated subprograms as methods of protected types, and declaring instances of
subprograms as methods. Since most reasonable use cases for doing this involve use of
generic action procedures, we will defer further consideration to Section 1.5, where we
introduce generic subprograms.

VHDL allows us to overload subprograms, and uses the parameter and result type
profiles to distinguish among them based on the types of parameters in a call. Where we
need to name a subprogram other than in a call, we can write a signature to indicate
which overloaded version we mean. The signature lists the parameter types and, for
functions, the return type, all enclosed in square brackets. This information is sufficient
to distinguish one version of an overloaded subprogram from other versions. We can use
a signature in attribute specifications, attribute names, and alias declarations. Subprogram
instantiations, introduced in VHDL-2008, are a further place in which we name a subprogram. If the uninstantiated subprogram is overloaded, we can include a signature in an
instantiation to indicate which uninstantiated version we mean. In such cases, the uninstantiated subprograms typically have one or more parameters of a formal generic type.
We use the formal generic type name in the signature. For example, if we have two uninstantiated subprograms declared as

```
procedure combine
  generic   ( type T )
  parameter ( x : T; value : bit );
```

```
procedure combine
  generic   ( type T )
  parameter ( x : T; value : integer );
```

the procedure name **combine** is overloaded. We can use a signature in an instantiation as follows:

```
procedure combine_vec_with_bit is new combine[T, bit]
  generic map ( T => bit_vector );
```

VHDL-2008 specifies that a formal generic type name of an uninstantiated subprogram is made visible within a signature in an instantiation of the subprogram. Thus, in this example, the signature distinguishes between the two uninstantiated subprograms, since only one of them has a profile with T for the first parameter and **bit** for the second. The T in the signature refers to the formal generic type for that version of the subprogram.

As with packages, we can also include a generic map in a subprogram, following the generic list. Such a subprogram is called a *generic-mapped subprogram*. A generic-mapped procedure has the form

```
procedure identifier
  generic ( ... )
  generic map ( ... )
  parameter ( ... ) is
  ...   -- declarations
begin
  ...   -- statements
end procedure identifier;
```

and a generic-mapped function has the form

```
function identifier
  generic ( ... )
  generic map ( ... )
  parameter ( ... ) return result_type is
  ...   -- declarations
begin
  ...   -- statements
end function identifier;
```

The generic list defines the generics, and the generic map aspect provides actual values and type for those generics. Like generic-mapped packages, we would not normally write a generic-mapped subprogram explicitly, since the feature is included in the language as a definitional aid. Hence, we won't dwell on them further, but simply mention them here to raise awareness in case an analyzer produces a seemingly cryptic error message.

1.5 Generic Subprograms

As well as generic constants and types, VHDL-2008 allows us to declare generic subprograms. We declare a formal generic subprogram in a generic list, representing some subprogram yet to be specified, and include calls to the formal generic subprogram within the unit that has the generic list. When we instantiate the unit, we supply an actual subprogram for that instance. Each call to the formal generic subprogram represents a call to the actual subprogram in the instance. The way we declare a formal generic subprogram is to write a subprogram specification in the generic list. The specification must be for a simple subprogram; that is, the subprogram must not contain a generic list itself.

We will illustrate formal generic subprograms with a number of examples based on typical use cases. One important use case is to supply an operation for use with a formal generic type declared in the same generic list as the subprogram. Recall, from our discussion in Section 1.1, that the only operations we can assume for a formal generic type are those defined for all actual types, such as assignment, equality and inequality. We can use a formal generic subprogram to explicitly provide further operations.

EXAMPLE 1.9 *Supplying an operator for use with a formal generic type*

In Example 1.2, we attempted to define a counter that could count with a variety of types. However, our attempt failed because we could not use the "+" operator to increment the count value. We can rectify this by declaring a formal generic function for incrementing the count value:

```vhdl
entity generic_counter is
  generic ( type      count_type;
            constant reset_value : count_type;
            function increment ( x : count_type )
                                 return count_type );
  port ( clk, reset : in  bit;
         data        : out count_type );
end entity generic_counter;
```

We can then use the increment function in the architecture:

```vhdl
architecture rtl of generic_counter is
begin
  count : process (clk) is
  begin
    if rising_edge(clk) then
      if reset = '1' then
        data <= reset_value;
      else
        data <= increment(data);
      end if;
    end if;
```

```
    end process count;
  end architecture rtl;
```

Having revised the counter in this way, we can instantiate it with various types. For example, to create a counter for **unsigned** values, we define a function, **add1**, to increment using the "+" operator on **unsigned** values and provide it as the actual for the **increment** generic.

```
use IEEE.numeric_std.all;
function add1 ( arg : unsigned ) return unsigned is
begin
  return arg + 1;
end function add1;

signal clk, reset : bit;
signal count_val  : unsigned(15 downto 0);
...

counter : entity work.generic_counter(rtl)
  generic map ( count_type  => unsigned(15 downto 0),
                reset_value => (others => '0'),
                increment   => add1 )  -- add1 is the
                                       -- actual function
  port map ( clk => clk, reset => reset, data => count_val );
```

In the instance, we specify a subtype of **unsigned** as the actual type for the formal generic type **count_type**. That subtype is then used as the subtype of the formal generic constant **reset_value** in the instance, so the actual value is a vector of 16 elements. The subtype is also used for the parameters of the formal generic function **increment** in the instance, so we must provide an actual function with a matching profile. The **add1** function meets that requirement, since it has **unsigned** as its parameter and result type. Within the instance, whenever the process calls the **increment** function, the actual function **add1** is called.

We can instantiate the same entity to create a counter for the **traffic_light_colour** type defined in Example 1.2. Again, we define a function, **next_color**, to increment a value of the type, and provide the function as the actual for the **increment** generic.

```
type traffic_light_color is (red, yellow, green);
function next_color ( arg  : traffic_light_color )
                      return traffic_light_color is
begin
  if arg = traffic_light_color'high then
    return traffic_light_color'low;
  else
    return traffic_light_color'succ(arg);
  end if;
end function next_color;
```

```
signal east_light : traffic_light_color;
...

east_counter : work.generic_counter(rtl)
  generic map ( count_type  => traffic_light_color,
                reset_value => red,
                increment   => next_color ) -- next_color is the
                                            -- actual function
  port map ( clk => clk, reset => reset, data => east_light );
```

When we declare a formal generic subprogram in a generic list, we can specify a default subprogram that is to be used in an instance if no actual generic subprogram is provided. The declaration is of the form

generic list (...;
                     subprogram_specification **is** *subprogram_name*;
                     ...);

The subprogram that we name must be visible at that point. It might be declared before the uninstantiated unit, or it can be another formal generic subprogram declared earlier in the same generic list. In the case of an uninstantiated package, we cannot name a subprogram declared in the package as a default subprogram, since items declared within the package are not visible before they are declared.

EXAMPLE 1.10 *Error reporting in a package*

Suppose we are developing a package defining operations to be used in a design and need to report errors that arise while performing operations. We can declare a formal generic procedure in the package to allow separate specification of the error-reporting action. We can also declare a default procedure that simply issues a report message. We need to declare the default action procedure separately from the package so that we can name it in the generic list. We will declare it in a utility package:

```
package error_utility_pkg is
  procedure report_error ( report_string   : string;
                           report_severity : severity_level );
end package error_utility_pkg;

package body error_utility_pkg is
  procedure report_error ( report_string   : string;
                           report_severity : severity_level ) is
  begin
    report report_string severity report_severity;
  end procedure report_error;
end package body error_utility_pkg;
```

We can now declare the operations package:

```
package operations is
  generic ( procedure error_action
                ( report_string  : string;
                  report_severity : severity_level )
                is work.error_utility_pkg.report_error );

  procedure step1 ( ... );
  ...

end package operations;

package body operations is

  procedure step1 ( ... ) is
  begin
    ...
    if something_is_wrong then
      error_action("Something is wrong in step1", error);
    end if;
    ...
  end procedure step1;
  ...

end package body operations;
```

If issuing a report message is sufficient for a given design, it can instantiate the operations package without providing an actual generic subprogram:

```
package reporting_operations is new work.operations;
use reporting_operations.all;
...

step1 ( ... );
```

If something goes wrong during execution of **step1** in this instance, the call to error_action results in a call to the default generic subprogram **report_error** defined in the utility package. Another design might need to log error messages to a file. The design can declare a procedure to deal with error messages as follows:

```
use std.textio.all;
file log_file : text open write_mode is "error.log";
procedure log_error ( report_string  : string;
                      report_severity : severity_level ) is
  variable L : line;
begin
  write(L, severity_level'image(report_severity));
```

```
    write(L, string'(": ");
    write(L, report_string);
    writeline(log_file, L);
end procedure log_error;
```

The design can then instantiate the operations package with this procedure as the actual generic procedure:

```
package logging_operations is new work.operations
  generic map ( error_action => log_error );
use logging_operations.all;
...

step1 ( ... );
```

In this instance, when something goes wrong in **step1**, the call to **error_action** results in a call to the procedure **log_error**, which writes the error details to the log file. Since the actual procedure is declared in the context of the instantiating design, it has access to items declared in that context, including the file object **log_file**. By providing this procedure as the actual generic procedure to the package instance, the instance is able to "import" that context via the actual procedure.

In many use cases where an operation is required for a formal generic type, there may be an overloaded version of the operation defined for the actual generic type at the point of instantiation. VHDL-2008 provides a way to indicate that the default for a generic subprogram is a subprogram, directly visible at the point of instantiation, with the same name as the formal generic subprogram and a matching profile. We use the box symbol ("<>") in place of a default subprogram name in the generic declaration. For example, we might write the following in a generic list of a package:

```
function minimum ( L, R : T ) return T is <>
```

If, when we instantiate the package, we omit an actual generic function, and there is a visible function named **minimum** with the required profile, then that function is used. Normally, the parameter type T used in the declaration of the formal generic subprogram is itself a formal generic type declared earlier in the generic list. We provide an actual type for T in the instance, and that determines the parameter type expected for the visible default subprogram. If we define the formal generic subprogram with the same name and similar profile to a predefined operation, we can often rely on a predefined operation being visible and appropriate for use as the default subprogram. We will illustrate this with an example.

EXAMPLE 1.11   *Dictionaries implemented as binary search trees*

The following package defines an abstract data type for dictionaries implemented as binary search trees. A dictionary contains elements that are each identified by a key value. The formal generic function **key_of** determines the key for a given element.

No default function is provided, so we must supply an actual function on instantiation of the package. The formal function "<" is used to compare key values. The default function is specified using the "<>" notation, so if an appropriate function named "<" is directly visible at the point of instantiation, we don't need to specify an actual function.

```
package dictionaries is
  generic ( type element_type;
            type key_type;
            function key_of ( E : element_type )
                              return key_type;
            function "<" ( L, R : key_type )
                          return boolean is <> );

  type dictionary_type;

  -- tree_record and structure of dictionary_type are private
  type tree_record is record
    left_subtree, right_subtree : dictionary_type;
    element : element_type;
  end record tree_record;
  type dictionary_type is access tree_record;

  procedure lookup ( dictionary : in dictionary_type;
                     lookup_key : in key_type;
                     element : out element_type;
                     found : out boolean );

  procedure search_and_insert ( dictionary : in dictionary_type;
                                element : in element_type;
                                already_present : out boolean );

end package dictionaries;
```

The package body is shown below, with the body of the **search_and_insert** procedure omitted for brevity.

```
package body dictionaries is

  procedure lookup ( dictionary : in dictionary_type;
                     lookup_key : in key_type;
                     element : out element_type;
                     found : out boolean ) is
    variable current_subtree : dictionary_type := dictionary;
  begin
    found := false;
    while current_subtree /= null loop
      if lookup_key < key_of( current_subtree.element ) then
```

```
            lookup ( current_subtree.left_subtree, lookup_key,
                     element, found );
          elsif key_of( current_subtree.element ) < lookup_key then
            lookup ( current_subtree.right_subtree, lookup_key,
                     element, found );
          else
            found := true;
            element := current_subtree.element;
            return;
          end if;
        end loop;
      end procedure lookup;

      procedure search_and_insert ( dictionary : in dictionary_type;
                                    element : in element_type;
                                    already_present : out boolean ) is
        ...

    end package body dictionaries;
```

In the function **lookup**, we use the formal generic function **key_of** to get the key for a candidate element in the dictionary. We compare the key with the value of the **lookup_key** parameter using the formal generic function "<".

Suppose we require a dictionary of test patterns that use time values as keys. We can instantiate the dictionaries package using our test-pattern type as the actual for **element_type** and **time** as the actual for **key_type**. We need to declare a function to get the time key for a test pattern:

```
type test_pattern_type is ...;

function test_id_of ( test_pattern : in test_pattern_type )
                    return time is
begin
  return ...;
end function test_id_of;
```

We don't need to define a function for use as the actual for the formal generic function "<". Since the predefined function "<" operating on **time** values is directly visible at the point of instantiation, it can be used implicitly as the actual function. As a result, the test patterns will be sorted into ascending order of time in the dictionary. We can write the package instantiation as:

```
package test_pattern_dictionaries is new work.dictionaries
  generic map ( element_type => test_pattern_type,
                key_type => time,
                key_of => test_id_of );
```

We can then call the operations defined in the instance:

```
use test_pattern_dictionaries.all;
variable test_set : dictionary_type;
variable generated_test, sought_test : test_pattern_type;
variable was_present : boolean;
...

search_and_insert ( test_set, generated_test, was_present );
assert not was_present
  report "Test at " & time'image(test_id_of(generated_test))
                   & " previously generated";
...
lookup ( test_set, 10 ns, sought_test, was_present );
assert was_present
  report "Test at 10 ns not found in test set";
```

EXAMPLE 1.12   *Dictionary traversal with an action procedure*

We can augment the dictionary abstract data type with an operation for traversing a
dictionary to apply an action to each element. We define the traversal procedure as
an uninstantiated procedure within the uninstantiated dictionaries package:

```
package dictionaries is
  generic ( ... );
  ...
  procedure traverse
    generic   ( procedure action ( element : in element_type ) )
    parameter ( dictionary : in dictionary_type );

end package dictionaries;

package body dictionaries is
  ...
  procedure traverse
    generic   ( procedure action ( element : in element_type ) )
    parameter ( dictionary : in dictionary_type ) is
  begin
    if dictionary = null then
      return;
    end if;
    traverse ( dictionary.left_subtree );
    action   ( dictionary.element );
    traverse ( dictionary.right_subtree );
  end procedure traverse;

end package body dictionaries;
```

Given this augmented package and the same instance as in Example 1.11, we can use the **traverse** procedure to count the number of elements in a dictionary. We first declare an action procedure:

```
variable test_pattern_count : natural := 0;
procedure count_a_test_pattern
            ( test_pattern : in test_pattern_type ) is
begin
  test_pattern_count := test_pattern_count + 1;
end procedure count_a_test_pattern;
```

We need to include the parameter, even though it is not used, since the profile of the action procedure must match that of the formal generic procedure. We instantiate the **traverse** procedure in the declarative part of the design:

```
procedure count_test_patterns is new traverse
  generic map ( action => count_a_test_pattern );
```

and then call the instance:

```
count_test_patterns(test_set);
assert test_pattern_count > 0
  report "The test patterns have gone missing!";
```

We can use a separate instantiation of the **traverse** procedure to perform a different action. For example, if we need to dump a list of test patterns to a file in order of their time, we would define an action procedure:

```
type test_pattern_file is file of test_pattern_type;
file dump_file : test_pattern_file;
procedure dump_a_test_pattern
            ( test_pattern : in test_pattern_type ) is
begin
  write(dump_file, test_pattern);
end procedure dump_a_test_pattern;
```

In this case, the parameter to the action procedure is used. We instantiate the **traverse** procedure in the declarative part of the design:

```
procedure dump_test_patterns is new traverse
  generic map ( action => dump_a_test_pattern );
```

and then call the instance:

```
file_open(dump_file, "test_patterns.dmp", write_mode);
dump_test_patterns(test_set);
file_close(dump_file);
```

The recursive **traverse** procedure in Example 1.12 further illustrates the rules we mentioned in Section 1.3 for interpreting names in uninstantiated units. The reference to the name **traverse** within that procedure is interpreted, in each instance of the procedure, as a reference to the instance. Thus each instance is properly recursive. This is the only situation where we can write a call to an uninstantiated subprogram.

In each of the examples we have seen, the subprogram that we provide as an actual, either explicitly or implicitly, for a formal generic subprogram has the same parameter and result type profile as the formal. In fact, the rule is stronger than that. The actual and formal subprograms must have *conforming profiles*, which means both are procedures or both are functions; the parameter and result type profiles of the two subprograms are the same; and corresponding parameters have the same class (**signal**, **variable**, **constant**, or **file**) and mode (**in**, **out**, or **inout**). The purpose of these rules is to ensure that a call to the formal subprogram will be legal for whatever actual subprogram is provided. As a counter example, suppose the formal subprogram had a signal parameter of a given type, and the actual subprogram had a variable parameter of the same type. A call to the formal subprogram would provide a signal as the actual parameter. However, the actual subprogram would expect a variable, and would perform variable assignments on it. This is clearly an error, even though the parameter and result type profiles of the two subprograms match. The additional requirements for profile conformance avoid this kind of error.

There are two further rules relating to the parameters of generic subprograms. The first is that, if a formal parameter of a formal generic subprogram has a default value, that value is used when an actual parameter is omitted, regardless of whether the corresponding formal parameter of the actual subprogram has a default value. An example will help clarify this. Suppose we declare an entity with a formal generic subprogram, and a corresponding architecture, as follows:

```
entity up_down_counter is
  generic ( type T;
            function add ( x : T; by : integer := 1 ) return T )
  port ( ... );
end entity up_down_counter;

architecture rtl of up_down_counter is
begin
  count : process (clk) is
  begin
    if rising_edge(clk) then
      if mode = "1" then
        count_value <= add(count_value); -- use default value
      else
        count_value <= add(count_value, -1);
      end if;
    end if;
  end process count;
end architecture rtl;
```

The formal generic subprogram **add** has a parameter **by** with the default value 1. In the first call to **add** within the architecture, we omit a value for **by**, so the default value 1 is used. This allows an analyzer to compile the call with the default value independently of any instantiation of the enclosing entity that we might write subsequently. For example, suppose we instantiate the entity with an actual generic subprogram declared as follows:

```
function add_int ( a : integer; incr : integer := 0 )
                return integer is
begin
  return a + incr;
end function add_int;
...


int_counter : entity work.up_down_counter(rtl)
  generic map ( T => integer; add => add_int )
  port map ( ... );
```

In this instance, the actual generic subprogram associated with **add** has the default value 0 for its second parameter. Despite this, the first call to the subprogram in the architecture still uses the default value 1 for the **by** parameter, since that is what is declared for the formal generic subprogram.

The rule dealing with default values for parameters also applies to the case where the parameter of the formal generic subprogram has no default value. In that case, a call must supply a value, even if the actual generic subprogram in an instance has a default value for the parameter. For example, in the **up_down_counter** entity, had we declared the formal generic function **add** as follows:

```
function add ( x : T; by : integer ) return T
```

the first call within the architecture would have to specify an actual value for the **by** parameter. The fact that the function **add_int** supplied as the actual generic subprogram in the instance has a default value for its second parameter cannot be used within the architecture.

The second rule relating to parameters of generic subprograms is that the parameter-subtype constraints of the actual subprogram apply when the subprogram is called, not the parameter-subtype constraints of the formal subprogram. To illustrate, suppose we instantiate the **up_down_counter** entity with a different function, as follows:

```
function add_nat ( a : natural; incr : natural := 0 )
                return natural is
begin
  return a + incr;
end function add_nat;
...


nat_counter : entity work.up_down_counter(rtl)
```

```
generic map ( T => natural; add => add_nat )
port map ( ... );
```

In this instance, the second parameter of the actual generic subprogram is of the base type **integer** with a range constraint requiring the value to be non-negative. The second call within the architecture provides the value −1 for the parameter. While this conforms to the constraint on the **by** parameter of the formal generic subprogram, it does not conform to the constraint on the corresponding parameter of the actual generic subprogram in the instance. Hence, when the function is called with that value in the instance, an error occurs.

1.5.1   Uninstantiated Methods in Protected Types

We now return to a discussion of the relationship between uninstantiated subprograms and protected types, mentioned in passing in Section 1.4. We build on our discussion of generic subprograms to provide motivating examples of the relationships that can occur. There are two cases to consider. The first is declaration of an instance of an uninstantiated subprogram as a method of a protected type, and the second is declaration of an uninstantiated subprogram within a protected type.

Starting with the first case, if we have an uninstantiated subprogram declared outside a protected type, and we declare an instance of the subprogram within the protected type declaration, the instance becomes a method of the protected type. The scheme is

```
procedure uninstantiated_name
  generic ( ... )
  parameter ( ... );

type PT is protected
  ...
  procedure instance_name is new uninstantiated_name
    generic map ( ... );
  ...
end protected PT;
```

We can declare a shared variable of the protected type and call the method:

```
shared variable SV : PT;
...

SV.instance_name ( ... );
```

On the face of it, there seems no purpose to this scheme. The uninstantiated subprogram, being outside the protected type, cannot refer to the items encapsulated within the protected type. So there would appear to be no reason for instantiating the subprogram in the protected type. However, we can provide controlled access to the encapsulated items via a method of the protected type provided as an actual generic subprogram to the instance. The refinement to the scheme is:

```
procedure uninstantiated_name
  generic ( ...; formal_generic_subprogram; ... )
  parameter ( ... );

type PT is protected
  method_declaration;
  procedure instance_name is new uninstantiated_name
    generic map ( ..., method_name, ... );
  ...
end protected PT;
```

In this scheme, the method has access to the encapsulated items within the protected type. When the instance invokes the actual generic subprogram, the method is called.

EXAMPLE 1.13   *Test-vector set with tracing*

Suppose we have an uninstantiated subprogram that gets a test-vector value corresponding to a specified time and that writes the vector value to the standard output file. The procedure has a formal generic subprogram representing the action to perform to get the test vector.

```
procedure trace_test_vector is
  generic  ( impure function get_test_vector
                            ( vector_time : time )
                            return test_vector )
  parameter ( vector_time : time ) is
  variable vector : test_vector;
  use std.textio.all;
  variable L : line;
begin
  write(L, now);
  write(L, string'(": "));
  vector := get_test_vector(vector_time);
  ...  -- write test vector
  writeline(output, L);
end procedure trace;
```

We can declare a protected type representing a set of test vectors to be applied at various times. The protected type has a method for getting a test vector for a specific time. We include an instance of the **trace_test_vector** procedure as a method to trace a test vector from the particular set represented by a shared variable of the protected type. The protected type declaration is:

```
type test_set is protected
  ...
  impure function get_vector_for_time ( vector_time : time )
                            return test_vector;
```

```
procedure trace_for_time is new trace_test_vector
   generic map ( get_test_vector => get_vector_for_time );
```

```
end protected test_set;
```

We might declare two shared variables of this protected type, representing two distinct sets of test vectors:

```
shared variable main_test_set, extra_test_set : test_set;
```

If we invoke the **trace_for_time** method on one of the shared variables:

```
main_test_set.trace_for_time(100 ns);
```

the instance of the **trace_test_vector** procedure invokes the actual subprogram provided for the instance of the protected type. That is, it invokes the **get_vector_for_time** method associated with the shared variable **main_test_set**. If, on the other hand, we invoke the **trace_for_time** method on the other shared variable:

```
extra_test_set.trace_for_time(100 ns);
```

the instance of the **trace_test_vector** procedure invokes the **get_vector_for_time** method associated with the shared variable **extra_test_set**. What this reveals is that each shared variable of the protected type binds its **get_vector_for_time** method, which has access to the shared variable's state, as the actual generic procedure in its instance of the **trace_test_vector** procedure. That instance, provided as a method of the shared variable, thus has indirect access to the shared variable's state.

The second case to consider is declaration of an uninstantiated subprogram within a protected type. That uninstantiated procedure is not itself a method, since it cannot be called. However, it can be instantiated within the protected type to provide a method. Moreover, each shared variable of the protected type contains a declaration of the uninstantiated subprogram. That subprogram can be instantiated, giving a subprogram that has access to the items encapsulated in the shared variable. We will illustrate these mechanisms with an example.

EXAMPLE 1.14   *Stimulus list with visitor traversal*

For a design requiring **signed** stimulus values, we can declare a procedure for displaying a **signed** value to the standard output file, as follows:

```
procedure output_signed ( value : in signed ) is
   use std.textio.all;
   variable L : line;
begin
   write(L, value);
```

```
    writeline(output, L);
end procedure output_signed;
```

We also declare a protected type for a list of **signed** stimulus values:

```
type signed_stimulus_list is protected
  ...

  procedure traverse_with_in_parameter
    generic ( procedure visit ( param : in signed ) );

  procedure output_all is new traverse_with_in_parameter
    generic map ( visit => output_signed );

end protected signed_stimulus_list;
```

The protected type includes an uninstantiated procedure to apply a visitor procedure to each element in the list of **signed** values. It instantiates the traversal procedure to provide a method that displays each element. We can use this protected type to declare a shared variable and then invoke the method to display its element values:

```
shared variable list1 : signed_stimulus_list;
...


list1.output_all;
```

Suppose now we want to use the traversal procedure to accumulate the sum of element in a list so that we can calculate the average value. We can provide another action procedure and use it in a further instantiation of the traversal procedure:

```
variable sum, average : signed(31 downto 0);
variable count : natural := 0;

procedure accumulate_signed ( value : in signed ) is
begin
  sum := sum + value;
  count := count + 1;
end procedure accumulate_signed;

procedure accumulate_all_list1 is
  new list1.traverse_with_in_parameter
    generic map ( visit => accumulate_signed );
...


accumulate_all_list1;
average := sum / count;
```

In this case, the instance is a procedure declared externally to the protected type. However, since it is an instance of a subprogram defined within the shared variable list1, the instance has access to the encapsulated items within list1. The instance **accumulate_all_list1** thus applies the **accumulate_signed** visitor procedure to each element within list1.

If we want to calculate the average value of any list of elements, we need to wrap these declarations up in a procedure that has a shared variable as a parameter. That includes declaring the instance of the traversal procedure within the outer procedure. The complete procedure would be:

```
procedure calculate_average
  ( variable list : inout signed_stimulus_list
    variable average : out signed ) is

  variable sum : signed(average'range);
  variable count : natural := 0;

  procedure accumulate_signed ( value : in signed ) is
  begin
    sum := sum + value;
    count := count + 1;
  end procedure accumulate_signed;

  procedure accumulate_all is
    new list.traverse_with_in_parameter
      generic map ( visit => accumulate_signed );

begin
  accumulate_all;
  average := sum / count;
end procedure calculate_average;
```

In this case, the instance of the traversal procedure is also declared externally to the protected type. However, it is an instance of the subprogram defined within the shared variable list provided as a parameter to the **calculate_average** procedure. Logically, each time the **calculate_average** procedure is called, a new instance of the traversal procedure is defined particular to the actual shared variable provided as the parameter. The instance thus applies the local **accumulate_signed** visitor procedure to each element within the actual shared variable.

1.6   Generic Packages

One of the common uses of packages is to declare an abstract data type (ADT), consisting of a named type and a collection of operations on values of the type. We have seen in Section 1.2 that we can include a generic list in a package declaration to make the package reusable for different actual types and operations. Often, the package for an ADT is reusable in this way.

Suppose we have an ADT specified in a package with generics, and we want to provide a further package extending the types and operations of the ADT. To make the extension package reusable, we would have to provide a generic type to specify an instance of the ADT named type, along with generic subprograms for each of the ADT operations. If the ADT has many operations, specifying them as actual generic subprograms in every instance of the extension package would be extremely onerous. To avoid this, VHDL-2008 allows us to specify an instance of the ADT package as a *formal generic package* of the extension package. Once we've instantiated the ADT package, we then provide that instance as the *actual generic package* of the extension package.

There are three forms of formal generic package declaration that we can write in a generic list. The first form is:

```
generic ( ...;
          package formal_pkg_name is new uninstantiated_pkg_name
            generic map ( <> );
          ... );
```

In this case, **formal_pkg_name** represents an instance of the **uninstantiated_pkg_name** package, for use within the enclosing unit containing the generic list. In most use cases, the enclosing unit is itself an uninstantiated package. However, we can also specify formal generic packages in the generic lists of entities and subprograms. When we instantiate the enclosing unit, we provide an actual package corresponding to the formal generic package. The actual package must be an instance of the named uninstantiated packge. The box notation "<>" written in the generic map of the formal generic package specifies that the actual package is allowed to be any instance of the named uninstantiated package. We use this form when the enclosing unit does not depend on the particular actual generics defined for the actual generic package.

No doubt, all of this discussion of packages within packages and generics at different levels can become confusing. The best way to motivate the need for formal generic packages and to sort out the relationships between the pieces is with an example.

EXAMPLE 1.15 *Fixed-point complex numbers*

VHDL-2008 defines a new package, **fixed_generic_pkg** (described in Section 8.4), for fixed-point numbers represented as vectors of **std_logic** elements. The package is an uninstantiated package, with generic constants specifying how to round results, how to handle overflow, the number of guard bits for maintaining precision, and whether to issue warnings. The package defines types **ufixed** and **sfixed** for unsigned and signed fixed-point numbers; and numerous arithmetic, conversion and input/output operations. We can instantiate the package with values for the actual generic constants to get a version with the appropriate behavior for our specific design needs.

Now suppose we wish to build upon the fixed-point package to define fixed-point complex numbers, represented in Cartesian form with fixed-point real and imaginary parts. We want the two parts of a complex number to have the same left and right index bounds, implying the same range and precision for the two parts. We can achieve that constraint by defining the complex-number type and operations in a package with formal generic constants for the index bounds. The complex-number type is defined using the **sfixed** type from an instance of the fixed-point package, and the complex-number operations need to use fixed-point operations from that instance. Thus, we include a formal generic package in the generic list of the complex-number package, as follows:

```
library IEEE;
package complex_generic_pkg is
  generic ( left, right : integer;
             package fixed_pkg_for_complex is
                new IEEE.fixed_generic_pkg
                  generic map (<>) );

  use fixed_pkg_for_complex.all;

  type complex is record
    re, im : sfixed(left downto right);
  end record;

  function "-"  ( z : complex ) return complex;
  function conj ( z : complex ) return complex;
  function "+"  ( l : complex;  r : complex ) return complex;
  function "-"  ( l : complex;  r : complex ) return complex;
  function "*"  ( l : complex;  r : complex ) return complex;
  function "/"  ( l : complex;  r : complex ) return complex;

end package complex_generic_pkg;
```

Within the **complex_generic_pkg** package, the formal generic package **fixed_pkg_for_complex** represents an instance of the **fixed_generic_pkg** package. The box notation in the generic map indicates that any instance of **fixed_generic_pkg** will be appropriate as an actual package. The use clause makes items defined in the **fixed_pkg_for_complex** instance visible, so that **sfixed** can be used in the declaration of type **complex**. The generic constants **left** and **right** are used to specify the index bounds of the two record elements. The operations defined for **sfixed** in the **fixed_pkg_for_complex** instance are also used to implement the complex-number operations in the package body for **complex_generic_pkg**, as follows:

```
package body fixed_complex_pkg is
  function "-" ( z : complex ) return complex is
  begin
    return ( -z.re, -z.im );
  end function "-";
    ...
end package body fixed_complex_pkg;
```

In the "–" operation for type **complex**, the "–" operation for type **sfixed** is applied to each of the real and imaginary parts. The other operations use the **sfixed** operations similarly.

In a design, we can instantiate both the fixed-point package and the complex-number package according to our design needs, for example:

```
package dsp_fixed_pkg is new IEEE.fixed_generic_pkg
  generic map ( fixed_rounding_style => true,
                fixed_overflow_style => true,
                fixed_guard_bits => 3,
                no_warning => false );


package dsp_complex_pkg is new work.complex_generic_pkg
  generic map ( left => 3, right => -12,
                fixed_pkg_for_complex => dsp_fixed_pkg );
```

The first instantiation defines an instance of the fixed-point package, which provides the type **sfixed** and operations with the required behavior. The second instantiation defines an instance of the complex-number package with left and right bounds of 3 and –12 for the real and imaginary parts. The type **sfixed** and the corresponding operations used within this instance of the complex-number package are provided by the actual generic package **dsp_fixed_pkg**. We can use the packages to declare variables and apply operations as follows:

```
use dsp_fixed_pkg.all, dsp_complex_pkg.all;
variable a, b, z : complex
variable c : sfixed;
...

z := a + conj(b);
z := (c * z.re, c * z.im);
```

The second form of formal generic package that we can write in a generic list is:

```
generic ( ...;
          package formal_pkg_name is new uninstantiated_pkg_name
            generic map ( actual_generics );
          ... );
```

Again, *formal_pkg_name* represents an instance of the *uninstantiated_pkg_name* package, for use within the enclosing unit containing the generic list. The actual generics provided in the generic map of the formal generic package specify that the actual package must be an instance of the named uninstantiated package with those same actual generics. We generally use this form when the enclosing unit also has another formal generic package defined earlier in its generic list. The latter generic is expected to have a generic package that is the same instance as the actual for the earlier generic package. No doubt that statement is unfathomable due to the packages within packages within packages. An example, building on Example 1.15, will help to motivate the need for the language feature and show how it may be used.

EXAMPLE 1.16   *Mathematical operations on fixed-point complex numbers*

In Example 1.15, we defined a package for complex number that provided a complex-number type and basic arithmetic operations. We can build upon this package to define a further package for more advanced mathematical operations on complex values. We will also use a package of advanced mathematical operations defined for fixed-point values:

```
package fixed_math_ops is
  generic ( package fixed_pkg_for_math is
             new IEEE.fixed_generic_pkg
               generic map (<>) );

  use fixed_pkg_for_math.all;

  function sqrt ( x : sfixed ) return sfixed;
  function exp ( x : sfixed ) return sfixed;
  ...

end package fixed_math_ops;
```

This package has a formal generic package for an instance of the **fixed_generic_pkg** package, since the operations it applies to the function parameters of type **sfixed** must be performed using the behavior defined for the **sfixed** type in the package instance proving the type. This is a similar scenario to that described in Example 1.15.

The advanced complex-number operations must be performed using the same **sfixed** type and basic fixed-point operations used to define the complex-number type and operations. It must also use the advanced fixed-point operations and the complex-number type and operations, with those types and operations being based on the same **sfixed** type and basic fixed-point operations. Thus, the advance complex-number package must have formal generic packages for the fixed-point package, the fixed-point mathematical operations package, and the complex-number package, as follows:

```
package complex_math_ops is
  generic ( left, right : integer;
           package fixed_pkg_for_complex_math is
             new IEEE.fixed_generic_pkg
               generic map (<>);
           package fixed_math_ops is
             new work.fixed_math_ops
               generic map ( fixed_pkg_for_math =>
                             fixed_pkg_for_complex_math );
```

```
            package complex_pkg is
              new work.complex_generic_pkg
                generic map ( left => left, right => right,
                              fixed_pkg_for_complex =>
                                fixed_pkg_for_complex_math ) );

    use fixed_pkg_for_complex_math.all,
        fixed_math_ops.all, complex_pkg.all;

    function "abs" ( z : complex ) return sfixed;
    function arg   ( z : complex ) return sfixed;
    function sqrt  ( z : complex ) return complex;
    ...

end package complex_math_ops;
```

The package body is

```
package body complex_math_ops is

  function "abs" ( z : complex ) return sfixed is
  begin
    return sqrt(z.re * z.re + z.im * z.im);
  end function "abs";
  ...

end package body complex_math_ops;
```

We can now instantiate the packages for a given design. For example, given the instances **dsp_fixed_pkg** and **dsp_complex_pkg** declared in Example 1.15, we can also declare instances of the advanced fixed-point operations package and the advanced complex operations package:

```
package dsp_fixed_math_ops is new work.fixed_math_ops
  generic map ( fixed_pkg_for_math => dsp_fixed_pkg );

package dsp_complex_math_ops is new work.complex_math_ops
  generic map ( left => 3, right => -12,
                fixed_pkg_for_complex_math => dsp_fixed_pkg,
                fixed_math_ops => dsp_fixed_math_ops,
                complex_pkg     => dsp_complex_pkg );
```

The third form of formal generic package that we can write in a generic list is:

```
generic ( ...;
        package formal_pkg_name is new uninstantiated_pkg_name
          generic map ( default );
        ... );
```

This form is similar in usage to the second form, but replaces the actual generics with the reserved word **default**. We can use this third form when the named uninstantiated package has defaults for all of its formal generics. The actual package must then be an instance of the named uninstantiated package with all of the actual generics being the same as the defaults. Those actual generics (for the actual generic package) can be either explicitly specified when the actual package is instantiated, or they can be implied by leaving the actual generics unassociated. Thus, this third form is really just a notational convenience, as it saves us writing out the defaults again as actual generics in the generic map of the formal generic package.

While generic packages might seem to be rather complex to put into practice, we envisage that most of the time packages using generic packages will be developed by personnel in support of design teams. They would normally provide source code templates for designers to instantiate the packages, including instantiating any dependent packages as actual generics. Thus, the designers would be largely insulated from the complexity.

For the developers of such packages, however, there are a number of rules relating to formal and actual generic packages. As we have mentioned, the actual package corresponding to a formal generic package must be an instance of the named uninstantiated package. To summarize the rules relating to the generic map in the formal generic package:

- If the generic map of the formal generic package uses the box ("<>") symbol, the actual generic package can be any instance of the named uninstantiated package.

- If the formal generic package declaration includes a generic map with actual generics, then the actual generics in the actual package's instantiation must match the actual generics in the formal generic package declaration.

- If the formal generic package declaration includes a generic map with the reserved word **default**, then the actual generics in the actual package's instantiation must match the default generics in the generic list of the named uninstantiated package.

The meaning of the term "match" applied to actual generics depends on what kind of generics are being matched. For generic constants, the actuals must be the same value. It doesn't matter whether that value is specified as a literal, a named constant, or any other expression. For a generic type, the actuals must denote the same subtype; that is, they must denote the same base type and the same constraints. Constraints on a subtype include range constraints, index ranges and directions, and element subtypes. For generic subprograms, the actuals must refer to the same subprogram, and for generic packages, the actuals must refer to the same instance of a specified uninstantiated package.

In the case of a default generic subprogram implied by a box symbol in the generic list of the named uninstantiated package, the actual subprogram must be the subprogram of the same name and conforming profile directly visible at the point where the formal generic package is declared. For example, if an uninstantiated package is declared as

```
package pkg1 is
  generic ( function "<" ( L, R : integer )
                    return boolean is <> ) );
```

```
  ...
end package pkg1;
```

we can declare a second package as follows:

```
package pkg2 is
  generic ( package inst1 is new pkg1 generic map ( default ) );
  ...
end package pkg2;
```

In this case, any package provided as an actual for inst1 must be an instance of pkg1, such as the following:

```
package ascending_pkg1 is new pkg1
  generic map ( T => integer );
```

Since the predefined "<" function for integer is visible at the point of declaring ascending_pkg1, that function is used as the actual for the generic function "<" in the instance of pkg1. At the place of declaring the formal generic package inst1 within the generic list of pkg2, the predefined "<" function for integer is also directly visible, so it is this function that must be matched as the actual for "<" in any instance of pkg1 supplied as an actual for inst1. Thus, the following instantiation of pgk2 is legal:

```
package integer_pkg2 is new pkg2
  generic map ( inst1 => ascending_pkg1 );
```

1.7 Use Case: Generic Memories

In this use case, we will explore the use of extended generics for modeling memories. We will develop a package of operations on memories, with the memory address width and depth specified by generic constants, and the address and data types specified using generic types. The package declares a type for signals representing RAM storage, and operations to read, write, load and dump RAM contents. The package declaration is:

```
library IEEE;
use IEEE.std_logic_1164.std_logic_vector;

package memories is
  generic ( width : positive;
            depth : positive;
            type address_type;
            type data_type;
            pure function to_integer (a : address_type)
              return natural is <>;
            pure function to_address_type (a : natural)
              return address_type is <>;
            pure function to_std_logic_vector (d : data_type)
              return std_logic_vector is <>;
```

```vhdl
        pure function to_data_type (d : std_logic_vector)
          return data_type is <> );

type RAM_type is array (0 to 2**depth - 1) of data_type;

procedure read_RAM (signal    RAM      : in  RAM_type;
                    constant address : in  address_type;
                    signal   data    : out data_type);

procedure write_RAM (signal    RAM      : out RAM_type;
                     constant address : in  address_type;
                     constant data    : in  data_type);


type format_type is (binary, hex);

procedure load_RAM (signal   RAM            : out RAM_type;
                    constant file_name      : in  string;
                    constant format         : in  format_type;
                    constant start_address  : in  address_type
                      := to_address_type(0);
                    constant finish_address : in  address_type
                      := to_address_type(2**depth - 1);
                    variable ok             : out boolean);

procedure dump_RAM (signal   RAM            : in  RAM_type;
                    constant file_name      : in  string;
                    constant format         : in  format_type;
                    constant start_address  : in  address_type
                      := to_address_type(0);
                    constant finish_address : in  address_type
                      := to_address_type(2**depth - 1);
                    variable ok             : out boolean);


end package memories;
```

The formal generic constants **width** and **depth** specify the bit width of memory data and addresses, respectively. The formal generic types **address_type** and **data_type** are used for memory addresses and data, respectively. The memory has 2^{depth} locations, indexed from 0 to $2^{depth-1}$, each storing a **data_type** value. Since we need to use integer values to index an array storing the memory contents, we need a function to convert an address to an integer; hence, the formal generic function **to_integer**. We also specify formal generic functions for use in the load and dump operations: to convert from an integer to an **address_type** value, to convert from a **std_logic_vector** value to a **data_type** value, and to convert from a **data_type** value to a **std_logic_vector** value. The reason for the last two is that the load and dump operations will read and write data values using the same formatting as that used for **std_logic_vector** values.

The type **RAM_type** is an array type used in models for signals representing RAM contents. The procedures **read_RAM** and **write_RAM** each have a signal parameter of this type, as well as parameters for the address and data. The **load_RAM** and **dump_RAM** procedures also have a **RAM_type** signal parameter, and load from or store to a file whose name is specified in the **file_name** parameter. The start and finish addresses are specified as parameters, with default values specified as integers converted to **address_type** values using the formal generic conversion functions. The **ok** parameter indicates whether the operation was successful.

The package body is:

```
package body memories is

  procedure read_RAM (signal    RAM      : in  RAM_type;
                      constant address : in  address_type;
                      signal    data     : out data_type) is
  begin
    assert to_integer(address) <= 2**depth - 1;
    data <= RAM(to_integer(address));
  end procedure read_RAM;

  procedure write_RAM (signal    RAM     : out RAM_type;
                       constant address : in  address_type;
                       constant data    : in  data_type) is
  begin
    assert to_integer(address) <= 2**depth - 1;
    RAM(to_integer(address)) <= data;
  end procedure write_RAM;

  use std.textio.all;

  procedure load_RAM (signal    RAM             : out RAM_type;
                      constant file_name        : in  string;
                      constant format           : in  format_type;
                      constant start_address    : in  address_type
                        := to_address_type(0);
                      constant finish_address   : in  address_type
                        := to_address_type(2**depth - 1);
                      variable ok               : out boolean) is
    file load_file  : text;
    variable status : file_open_status;
  begin
    ok := false;
    file_open(f => load_file, external_name => file_name,
              open_kind => read_mode, status => status);
    if status /= open_ok then
      return;
    end if;
```

```
    -- code to read and parse memory file contents
    ...
    file_close(f => load_file);
    ok := true;
  end procedure load_RAM;

  procedure dump_RAM (signal    RAM              : in   RAM_type;
                      constant file_name         : in   string;
                      constant format            : in   format_type;
                      constant start_address     : in   address_type
                         := to_address_type(0);
                      constant finish_address    : in   address_type
                         := to_address_type(2**depth - 1);
                      variable ok                : out  boolean) is
    file dump_file  : text;
    variable status : file_open_status;
  begin
    ok := false;
    file_open(f => dump_file, external_name => file_name,
              open_kind => write_mode, status => status);
    if status /= open_ok then
      return;
    end if;
    -- code to write memory file contents
    ...
    file_close(f => dump_file);
    ok := true;
  end procedure dump_RAM;

end package body memories;
```

The **read_RAM** procedure asserts that the address, converted to an integer value, lies within the index range of the memory array type. It uses the converted address to index the RAM signal, and assigns the indexed element value to the **data** signal parameter. The **write_RAM** procedure is similar, but updates the indexed element using the **data** parameter value.

The **load_RAM** procedure attempts to open the file named by the **file_name** parameter. If it succeeds, the procedure then reads the file contents. Assuming the file is in Verilog memory format, the data values are read as **std_logic_vector** values, each of the width specified by the **width** generic. These values are converted to **data_type** values using the formal generic conversion function **to_data_type**. The **dump_RAM** procedure is similar, but converts **data_type** values to **std_logic_vector** values using the **to_std_logic_vector** formal generic function.

There are several common cases for address and data types. Specifically, addresses are commonly of type **natural**, **std_logic_vector**, or **unsigned**; and data values are commonly of type **natural**, **std_logic_vector**, **unsigned**, or **signed**. In support of these common cases, we can define a package of conversion functions as follows:

```vhdl
library IEEE;
use IEEE.std_logic_1164.std_logic_vector;
use IEEE.numeric_std.unsigned;

package memories_support is
  generic ( width : positive;
            depth : positive );

  -- Conversions for common actual types for address_type:
  -- natural, std_logic_vector, unsigned

  pure function to_integer (a: natural) return natural;

  -- to_integer [std_logic_vector return natural]
  -- provided by ieee.numeric_std_unsigned

  -- to_integer [unsigned return natural]
  -- provided by ieee.numeric_std

  pure function to_address_type (a : natural) return natural;

  pure function to_address_type (a : natural)
                                    return std_logic_vector;

  pure function to_address_type (a : natural) return unsigned;

  -- Conversions for common actual types for data_type:
  -- natural, std_logic_vector, unsigned, signed

  pure function to_std_logic_vector (d : natural)
                                    return std_logic_vector;

  pure function to_std_logic_vector (d : std_logic_vector)
                                    return std_logic_vector;

  pure function to_std_logic_vector (d : unsigned)
                                    return std_logic_vector;

  pure function to_std_logic_vector (d : signed)
                                    return std_logic_vector;


  pure function to_data_type (d : std_logic_vector)
                             return natural;

  pure function to_data_type (d : std_logic_vector)
                             return std_logic_vector;
```

```vhdl
  pure function to_data_type (d : std_logic_vector)
                             return unsigned;

  pure function to_data_type (d : std_logic_vector)
                             return signed;

end package memories_support;
```

The package has **width** and **depth** generic constants, since these are needed to determine the vector length for conversions from integer values to vector values. No **to_integer** conversions are needed for **std_logic_vector** or **unsigned**, since these are provided by the **numeric_std_unsigned** and **numeric_std** packages, respectively. The function bodies are either identities or wrappers around type conversions or conversion functions, as shown in the corresponding package body:

```vhdl
package body memories_support is

  pure function to_integer (a: natural) return natural is
  begin
    return a;
  end function to_integer;

  pure function to_address_type (a : natural) return natural is
  begin
    return a;
  end function to_address_type;

  pure function to_address_type (a : natural)
                                 return std_logic_vector is
  begin
    return IEEE.numeric_std_unsigned.to_stdlogicvector(a,
                                                       depth);

  end function to_address_type;

  pure function to_address_type (a : natural) return unsigned is
  begin
    return IEEE.numeric_std.to_unsigned(a, depth);
  end function to_address_type;

  pure function to_std_logic_vector (d : natural)
                                     return std_logic_vector is
  begin
    return IEEE.numeric_std_unsigned(d, width);
  end function to_std_logic_vector;
```

```
RAM_proc : process (clk) is
begin
  if rising_edge(clk) then
    if en = '1' then
      if wr = '1' then
        write_RAM(RAM, addr, data_in);
        data_out <= data_in;
      else
        read_RAM(RAM, addr, data_out);
      end if;
    end if;
  end if;
end process RAM_proc;

end architecture rtl;
```

Since we're using address and data types that are catered for by the support package, we instantiate that package in the architecture body, providing the memory depth and width values as actual generic constants. The use clause for the instance makes all of the required conversion functions for the address and data types directly visible. Next, we instantiate the memories package, providing the memory depth and width values as actuals for the generic constants, and the address and data port subtypes as actuals for the **address_type** and **data_type** formal generic types. Since functions with the required names and signatures are directly visible, they are used as actuals for the formal generic functions. The use clause makes the type and the procedures from the package instance directly visible. We use **RAM_type** as the type in a signal declaration for the SSRAM storage. The **RAM_init** process checks the value of the **hex_file_name** generic, and if it is not empty, calls the **load_RAM** procedure to initialize the SSRAM contents. The **RAM_proc** process uses the **write_RAM** and **read_RAM** procedures to implement memory operations.

Chapter 2

Other Major Features

The enhancement of generics that we described in Chapter 1 is one of several major new features in VHDL. In this chapter, we highlight the other major features that bring significant new power to the language.

2.1 External Names

One of the characteristics of VHDL is that it allows a verification testbench to be written in the same language as the design to be verified. However, some aspects of earlier versions of VHDL make it hard to verify designs. In particular, the scope and visibility rules are intended to help us manage name spaces in complex designs by enforcing abstraction of interfaces and hiding of internal information. While they are good for a design in isolation, they can prevent a testbench from accessing items internal to a design. A testbench may need to monitor the state of internal signals, or force internal signals to particular values.

VHDL-2008 provides a new naming feature, *external names*, that allows us to write a testbench that accesses items not normally visible according to the hierarchical scope and visibility rules. An external name specifies a hierarchical path through the design hierarchy to reach a declared constant, shared variable, or signal. Thus, a testbench using an external name must have sufficient knowledge of the hierarchical structure of the design for the path to be valid. Validity of the external name is assumed during analysis of the testbench, and is checked during elaboration of the complete design hierarchy.

An external name is written in the following form:

```
<< class external_pathname : subtype >>
```

where the class is one of the object classes **constant**, **signal**, or **variable**; the external pathname is the hierarchical path; and the subtype specifies a view of the object, in a way similar to an alias. As an example, a testbench might use the following external name to monitor the value of a signal within a design under verification:

```
assert <<signal .tb.duv.controller.state :
                std_logic_vector(0 to 4)>> /= "00000"
  report "Illegal controller state";
```

Within the testbench, this external name is a reference to a signal nested within the component labeled **controller**, which is nested within the component labeled **duv**, which is within the top-level entity **tb**. The signal is interpreted as a **std_logic_vector** indexed from 0 to 4. When the testbench is analyzed, the existence and type of the signal is not checked. However, once the complete design hierarchy is elaborated, the signal must exist and be of an appropriate type to match the subtype specified in the external name.

An external name is just a new form of name for a constant, signal or variable, so we can use an external name at any place where a name is appropriate, subject to some rules that we will return to shortly. That means we can refer to a constant or signal value in an expression, and we can assign to a signal or include it in a port map. The rules for forming a pathname only allow us to refer to items declared in concurrent regions of a design (packages, entities, architectures, blocks and generate statements), so an external variable name can only refer to a shared variable. We can use an external variable name to invoke a method of the shared variable, for example:

```
<<variable .tb.duv_behavior.msg_fifo :
           fifo_type>>.put(corrupt_msg);
```

For an external name that refers to an object of a composite type, we can refer to an element of the object. For example, given an array signal declared within a design, we can index the array with an external name as the prefix:

```
<<signal .tb.duv_rtl.data_bus :
         std_logic_vector(0 to 15)>>(8) <= '1';
```

One common use case is to declare an alias for an external name. If we do that, we need only write the full external name in the alias declaration. Thereafter, we can just use the shorter alias name, making the model more succinct. For example, if we need to refer to the **data_bus** signal in several places in a testbench, we could declare an alias for it:

```
alias duv_data_bus is
  <<signal .tb.duv_rtl.data_bus : std_logic_vector(0 to 15)>>;
```

and then just use the alias in the assignment and other places:

```
duv_data_bus(8) <= '1';
sign <= duv_data_bus(0);
```

In an alias declaration, we have the option of specifying a subtype after the alias name, giving us a view of the named object as being of that subtype, for example:

```
alias identifier : subtype is name;
```

However, when the name we are aliasing is an external name, the subtype is specified in the external name. We do not repeat the subtype (or specify a conflicting subtype!) after the alias name. So the following two alias declarations are illegal:

```
alias duv_data_bus : std_logic_vector(0 to 15) is -- illegal!
  <<signal .tb.duv_rtl.data_bus : std_logic_vector(0 to 15)>>;

alias duv_data_bus : std_logic_vector(15 downto 0) is -- illegal!
  <<signal .tb.duv_rtl.data_bus : std_logic_vector(0 to 15)>>;
```

We can use an external constant name (or an alias of such a name) in an expression, provided the constant has been elaborated and given a value by the time the expression is evaluated. In some cases, expressions are evaluated during elaboration of a design. For example, initial-value expressions and index-bound expressions in declarations are evaluated when the declaration is elaborated, so an external constant name appearing in those places must refer to a constant that has already been elaborated. We can ensure this is the case by writing the part of the design that includes the constant declaration prior to the part of the design that contains the external constant name. VHDL's elaboration rules specify that the design is elaborated in depth-first top-to-bottom order. To illustrate how we can take account of this order, suppose we have an entity and architecture for a design that declares a constant, as follows:

```
entity design is
  port ( ... );
end entity design;

architecture rtl is
  constant width : natural := 32;
  ...
begin
  ...
end architecture rtl;
```

Suppose also that we have a testbench entity and architecture:

```
entity testbench is
end entity testbench;

architecture directed of testbench is
  signal test_in :
    bit_vector(0 to <<constant .top.duv.width : natural>> - 1);
  ...
begin
  ...
end architecture directed;
```

We now assemble the design and testbench in a top-level entity and architecture:

```
entity top is
end entity top;

architecture level of top is
begin
  assert false
    "Width = " &
    integer'image(<<constant .top.duv.width : natural>>);
  duv : entity work.design(rtl);
```

```
  tb   : entity work.testbench(directed);
end architecture level;
```

In this case, the instance of the design under verification is elaborated before the testbench instance. Thus, the constant declaration is elaborated and given a value before the external constant name within the **tb** instance is elaborated. Had we written the two instances in the reverse order, the constant would not have been elaborated at the time of elaborating the external constant name, and an error would occur. The external constant name in the assertion statement, on the other hand, is not evaluated until the model is executed, by which time the model is completely elaborated. Thus, the external constant name is allowed to precede the instance of the design under test in which the constant is declared.

VHDL-2008 has a related rule regarding elaboration of a signal referenced by an external signal name. If such a name (or an alias of such a name) is used in a port map, the signal declaration must have been previously elaborated. The reason is that the hierarchy of signal nets and drivers is built during elaboration. If a signal used in a port map is not yet elaborated, the elaborator would have to revisit elaboration of that part of the design hierarchy once the signal declaration was encountered. In general, allowing such use of external signal names would make elaboration of signal nets indefinitely complicated. The rule preventing such use allows elaboration to proceed in a well-defined order, and is not onerous in practice. It usually just requires that the component instance in which the signal is declared be written before the instance referencing the signal in a port map. The typical scenario is that a design under verification be instantiated before the testbench code containing external names.

The pathname in an external name identifies the location of the referenced object within the design hierarchy. A design hierarchy has an instance of an entity and some associated architecture at the top level. The entity and architecture can contain declarations of objects. We can identify such an object by naming the entity followed by the object name. The entity and architecture can also contain nested regions, which can in turn contain declarations of objects. We can identify an object in such a nested region by joining the object name onto the name for the region and the name for the top-level entity. In the case of a block, the name for the region is the block label. In the case of a generate statement, the name for the region is the generate label. A for-generate also requires a value to indicate which iteration of the generate to use. In the case of a local package (see Section 1.3), the name for the region is the package name. Note that we cannot use the name of an uninstantiated package (see Section 1.2) in this way; we can only use the name of an instance of the uninstantiated package. In the case of a component instance, the name of the region is the component instance label, and the region is that corresponding to the bound entity and architecture. We can apply these rules recursively to build up a chain of region names, starting from the entity at the top of the design hierarchy and leading through levels of nesting to identify any object instantiated within the hierarchy.

The pathnames in the preceding external names are all examples of *absolute pathnames*, which specify the full chain of region names, starting from the top of the design hierarchy containing the external name. An absolute pathname starts with a dot symbol and separates each region name within the pathname with further dot symbols. The pathname ends with the simple name of the referenced object. Thus, the absolute path-name

.tb.duv_rtl.data_bus

refers to the object named **data_bus** declared within the entity and architecture bound to the component labeled **duv_rtl** within the top-level entity **tb**. Similarly, the absolute path-name

.tb.duv_rtl.memory(3).addr_bus

refers to the object named **addr_bus** within the for-generate iteration with index 3 within the component instance mentioned.

EXAMPLE 2.1   *Monitoring states in an embedded state machine*

Suppose we are verifying a system that includes a finite-state machine control unit embedded as a subcomponent. The control unit is described by the following entity and architecture:

```
library IEEE; use IEEE.std_logic_1164.all;
entity control is
  port ( clk, reset : in std_logic; ... );
end entity control;

architecture fsm of control is
  subtype state_type is std_logic_vector(3 downto 0);
  constant idle     : state_type := "0000";
  constant pending1 : state_type := "0001";
  ...
  signal current_state, next_state : state_type;
begin
  state_reg : process (clk) is ...
  fsm_logic : process (all) is ...
end architecture fsm;
```

Note, in passing, that the **fsm_logic** process uses the notation **all** in its sensitivity list, indicating that the process is sensitive to changes in all signals read within the process. This new VHDL-2008 feature is described in Section 6.2. The entity and architecture for the system being designed are:

```
library IEEE; use IEEE.std_logic_1164.all;
entity system is
  port ( clk, reset : in std_logic; ... );
end entity system;

architecture rtl of system is
  component control is
```

```
      port ( clk, reset : in std_logic; ... );
    end component control;
begin
  control_unit : component control
    port map ( clk => clk, reset => reset, ... );
    ...
end architecture rtl;
```

We can define a testbench entity and architecture that traces the sequence of states in the control unit, writing each to a file:

```
entity state_monitor is
  generic ( state_file_name : string );
end entity state_monitor;

architecture tracing of state_monitor is
  alias fsm_clk is
    <<signal .tb.system_duv.control_unit.clk : std_logic>>;
  alias fsm_state is
    <<signal .tb.system_duv.control_unit.current_state :
              std_logic_vector(3 downto 0)>>;
begin
  monitor : process (fsm_clk) is
    use std.textio.all;
    file state_file : text open write_mode is state_file_name;
  begin
    if falling_edge(fsm_clk) then
      write(L, fsm_state); writeline(state_file, L);
    end if;
  end process monitor;
end architecture tracing;
```

Note here that the external reference to the **clk** port of the **control_unit** instance treats the port as a signal declared in the region corresponding to the instance. This reflects the rule in VHDL that a port is a signal declared in the declarative region of an entity. A generic constant of an instance would similarly be referenced using an external constant name with a pathname for the instance.

The external references in this architecture assume that the complete design hierarchy has an entity named **tb** at the root, and that the instance of the system to be monitored is labeled **system_duv** within the top-level architecture. To satisfy those assumptions, we write the top-level entity and architecture as:

```
library IEEE; use IEEE.std_logic_1164.all;
entity tb is
end entity tb;

architecture monitoring of tb is
  signal system_clk, system_reset : std_logic;
```

```
    ...
  begin
    ... -- clock and reset generation

    system_duv : entity work.system(rtl)
      port map ( clk => system_clk, reset => system_reset, ... );

    state_monitor : entity work.state_monitor(tracing)
      generic map ( state_file_name => "fsm_states.dat" );

  end architecture monitoring;
```

Within the tracing architecture of the **state_monitor** entity, we write an external name for the **current_state** signal with a **std_logic_vector** subtype. Normally, we would declare an enumeration type for the states of a finite-state machine. If we declare such a type locally within the control unit architecture, it would not be visible to the external monitor. We would not be able to write an external name with an appropriate subtype for the referenced signal. That is why we used a **std_logic_ vector** subtype for the state type in this example. If we want to declare an enumeration type for an object that is to be externally monitored, we would have to declare the type in a package that is visible both in the object declaration and in the monitor.

In some testbenches, the testbench code is written in the same region as an instance of the design under verification. In those cases, there is no need to specify the absolute path starting from the top-level entity. Instead, we can use a *relative pathname*, consisting of the chain of region names starting from the immediately enclosing region, without the leading dot symbol. For example, if a testbench architecture includes an instance of the design under verification labeled **duv**, then the architecture could also contain the assertion statement:

```
  assert <<signal duv.controller.state :
                  std_logic_vector(0 to 4)>> /= "00000"
    report "Illegal controller state";
```

Since the starting point for the relative pathname is the enclosing architecture region, the first part of the pathname refers to the component instance, and subsequent parts refer to items nested within the bound entity and architecture.

An important point to note when we are talking about the innermost region for a relative pathname is that only concurrent regions are considered. If we write an external name with a relative pathname within a process or subprogram, that region does not count, since it is not a concurrent region. Moreover, if the name is within a package that is declared within a process or subprogram, the package region does not count either. We need to look outward in the design hierarchy to find an enclosing entity, architecture, block, or generate statement, or a package that is declared in such a region.

EXAMPLE 2.2   *Revised state monitoring for an embedded state machine*

Returning to the test bench of Example 2.1, we can write the state-monitoring code directly in the top-level architecture rather than in an instantiated entity and architecture. In that case, we can use relative pathnames, and so do not have to assume the name of the top-level entity. The revised top-level architecture is:

```
architecture monitoring of tb is
  signal system_clk, system_reset : std_logic;

  alias fsm_clk is
    <<signal system_duv.control_unit.clk : std_logic>>;
  ...
begin
  ... -- clock and reset generation

  system_duv : entity work.system(rtl)
    port map ( clk => system_clk, reset => system_reset, ... );

  monitor : process (fsm_clk) is
    use std.textio.all;
    file state_file : text open write_mode is state_file_name;
    alias fsm_state is
      <<signal system_duv.control_unit.current_state :
               std_logic_vector(3 downto 0)>>;
  begin
    if falling_edge(fsm_clk) then
      write(L, fsm_state); writeline(state_file, L);
    end if;
  end process monitor;

end architecture monitoring;
```

In this architecture, the alias declarations refer to external names identified with relative pathnames. The component label **system_duv** is declared in the same enclosing architecture region as the alias declarations, so that label is the one used in the pathnames. Even though the external name aliased to **fsm_state** is written within the process region, the innermost region considered is that of the enclosing architecture.

A further form of relative pathname allows us to identify an outer region as the starting point for the pathname. We write such a pathname using one or more leading "^" symbols in place of names, for example:

```
<<constant ^.^.comp.c : real>>
```

As for the relative pathname without the "∧" symbols, we initially start with the innermost concurrent region enclosing the external name. Then, for each "∧" symbol, we look in the next enclosing region. In the case of instantiated components, the region enclosing an instance of a bound entity and architecture is the region in which the instantiation is written. Thus, if we use this form of pathname in an entity or architecture, we are making a strong assumption about the context in which the entity and architecture are instantiated. Specifically, we are assuming that context also includes the names written in the pathname. The complete design hierarchy must be built in such a way as to ensure the assumption is met, otherwise an error will occur during elaboration.

EXAMPLE 2.3 *Relative pathname in a nested monitor*

Suppose we are verifying a multicore platform, in which each core includes an instance of a CPU described by the following entity and architecture.

```
entity CPU is
  ...
end entity CPU;

architecture BFM of CPU is
  use work.CPU_types.all;
  signal fetched_instruction : instruction_type;
  ...
begin
  ...
end architecture BFM;
```

The architecture includes a signal representing a fetched instruction. The multi-core platform is described by an entity with a generic constant specifying the number of cores. The architecture of the entity uses a for-generate statement to replicate instances of the CPU.

```
entity platform is
  generic ( num_cores : positive );
  port    ( ... );
end entity platform;

architecture BFM_multicore of platform is
  ...
begin
  cores : for core_num in 1 to num_cores generate
    processor : entity work.CPU(BFM) ...;
    ...
  end generate cores;
  ...
end architecture BFM_multicore;
```

We now consider the testbench that instantiates the platform entity and architecture. Again, we use a generic constant to determine the number of cores in the design under verification. We can include a monitor for each instantiated core by writing a for-generate statement in the testbench, mirroring that in the platform architecture.

```
entity testbench is
  generic ( num_cores : positive );
end entity testbench;

architecture test_BFM of testbench is
  ...
begin
  duv : entity work.platform(BFM_multicore)
    generic map ( num_cored => num_cores )
    port map    ( ... );

  monitors : for core_num in 1 to num_cores generate
    use work.CPU_types.all, work.CPU_trace.all;
    process is
    begin
      ...
      trace_instruction
        ( <<signal
          ^.duv.cores(core_num).processor.fetched_instruction :
            instruction_type>>,
          ... );
      ...
    end process;
  end generate monitors;
end architecture test_BFM;
```

The process within the generate statement includes an external name referring to the fetched_instruction signal in the corresponding core instance. The pathname uses the value of the **core_num** generate parameter to identify the corresponding iteration of the generate statement labeled **cores** in the design under verification. Since the external name is in a process nested within a generate statement, the generate statement region is the innermost region used as the starting point for the relative pathname. For that reason, the pathname starts with a "^" symbol to look outside the starting region to the enclosing architecture region. The **duv** component instance is declared in that region, so it can be used as the next part of the pathname.

We also need to be able to refer to an object declared in a package referenced by a design. For objects declared in the package declaration, we can just use the package name as a prefix in a normal selected name to refer to the object. However, objects declared in the package body are not visible to designs. They would normally be referenced indirectly using procedures declared in the package declaration. A testbench, on the other hand, can use an external name to refer to such a hidden object. An object in a package is not nested within the design hierarchy, but is considered to be nested within the library containing the package. So the chain of region names starts with the library logical name (the name defined by a library clause) and leads through the top-level package name and any nested package names to the referenced object.

A *package pathname* takes a similar form to an absolute pathname, but starts with an "@" symbol instead. That is followed by logical name of the library containing the package, then the package name, then the names of any intervening nested packages, and finally the object name. For example, given the following package declaration and body analyzed into the working library:

```
package p1 is
  . . .
end package p1;

package body p1 is
  . . .
  package p2 is
    signal s : bit;
  end package p2;
  . . .
end package body p1;
```

we could write the following external name to refer to the signal:

```
<<signal @work.p1.p2.s : bit>>
```

2.2   Force and Release

When verifying a design, we often would like to be able to override the value assigned to a signal in the normal course of design operation and force a different value onto the signal. One reason for doing this is to set up a test scenario by forcing values to a state that would normally be arrived at through a complex initialization sequence. Forcing the values allows us to bypass the sequence and set up the scenario quickly, and so reduce the verification time significantly. Another reason for forcing values is to inject erroneous values into the design to ensure that it detects the error or otherwise responds appropriately.

In earlier versions of VHDL, there was no way to override the value assigned to a signal by the design, other than using commands provided by a simulator. That meant we could not write VHDL testbench code to force signal values. VHDL-2008 now provides features for forcing and releasing the values of signals. We can force a signal with a *force assignment* of the form

```
signal_name <= force expression;
```

This is a sequential assignment written within a process forming part of the test-bench. The effect is to cause a delta cycle and to force the named signal to take on the value of the expression in that delta cycle, regardless of any value assigned to the signal by any normal signal assignment. The signal is considered to be active during the delta cycle, and if the forcing value is different from the previous value, an event occurs on the signal. Processes sensitive to changes on the signal value would then respond to the value change in the normal way.

The usual rules relating the type of the expression to the type of the target signal apply for force assignments. The target signal name can be a normal signal name, or it can be an external signal name or alias (see Section 2.1). Using external names would be a common use case, since we would often need to force an internal signal of a design from a testbench.

Once a signal has been forced, we can update the signal with another force assignment to change the overriding value, again causing the signal to become active and possibly to have another event. We can do this as often as needed. Eventually, if we want to stop forcing a signal, we can execute a *release assignment* of the form

```
signal_name <= release;
```

This causes a further delta cycle, with the signal being active. However, since the signal is no longer forced, the current values of its sources are used to determine the signal value in the normal way. We can think of this as the design "taking back control" of the signal.

EXAMPLE 2.4   *Simulated corruption of a state machine's state value*

Clocked sequential systems are usually controlled by a finite-state machine. If the storage for the current state is corrupted, the system may be able to recover by transitioning from the illegal state back to an initial state. A testbench can verify that a design under verification recovers correctly by forcing the signal storing the current state of the state machine to an illegal value. It can then release the signal and monitor recovery. The testbench process is:

```
verify_state_recovery : process is
  use work.control_pkg.all;
  alias clk is <<signal duv.clk : std_logic>>;
  alias current_state is
          <<signal duv.control.current_state : state_type>>;
begin
  ...
  -- inject corrupt state
  wait until falling_edge(clk);
  current_state <= force illegal_state_12;
  wait until falling_edge(clk);
  current_state <= release;
  -- monitor recovery activity
```

```
    . . .
end process verify_state_recovery;
```

Our discussion of force assignments has so far focused on signals. We can also force and release ports of a design, since they are a form of signal. However, for a port, we distinguish between the driving value and the effective value. The driving value is the value presented externally by an entity, and is determined by the internal sources within the entity. The effective value is the value seen internally by an entity and is determined by whatever is externally connected to the port, whether that be an explicitly declared signal or a port of an enclosing entity. Depending on the port mode and the external connections, the driving and effective values may be different. For example, an **inout**-mode port of type **std_logic** might drive a '0' value, but the externally connected signal might have another source driving a '1' value. In that case, the resolved value of the signal is 'X', and that value is seen as the effective value of the **inout**-mode port.

VHDL-2008 allows us to force the driving and effective values of a signal or port independently by including a *force mode* in an assignment. For explicitly declared signals, where the driving and effective values are the same, the distinction makes no difference. For ports and signal parameters, we can force the driving value by including the keyword **out** in the force assignment:

```
signal_name <= force out expression;
```

Alternatively, we force the effective value by including the keyword **in** in the force assignment:

```
signal_name <= force in expression;
```

Once we've forced a port's or signal parameter's driving value, we can stop forcing it by writing a release assignment with the keyword **out**:

```
signal_name <= release out;
```

Similarly, to release a forced effective value, we write a release assignment with the keyword **in**:

```
signal_name <= release in;
```

We can force and release driving values of ports of mode **out**, **inout**, and **buffer**, but not ports of mode **in**. Similarly, we can force and release driving values of signal parameters of mode **out** and **inout**, but not signal parameters of mode **in**. One of the VHDL-2008 changes for ports and parameters, described in Section 6.3, is that we can read the value of an **out**-mode port or parameter. This means that ports and signal parameters of all modes except **linkage** have effective values, and so we can force and release the effective value of a port or signal parameter of any mode except **linkage**.

If we omit the force mode (**out** or **in**) in a force or release assignment, a default force mode applies. For assignments to ports and signal parameters of mode **in** and to explicitly declared signals, the default force mode is **in**, forcing the effective value. For assignments to ports of mode **out**, **inout**, or **buffer**, and to signal parameters of mode **out** or **inout**, the default force mode is **out**, forcing the driving value.

EXAMPLE 2.5   *Forcing disconnection of a port's driving value*

Serial buses such as I^{2}C, USB and FireWire have bidirectional connections to the bus' physical wires. This allows a device to drive the clock and data wires when transmitting data and to sense the clock and data values when receiving. A testbench can model a broken data driver connection by forcing a 'Z' value on the output part of the bidirectional port, while allowing the input part of the port to operate normally. The code in the testbench is

```
...
-- Test scenario: break in the output connection
<<signal duv.SDA : std_logic>> <= force out 'Z';
-- Monitor device operation under this fault condition
...
-- Restore connection for the next scenario
<<signal duv.SDA : std_logic>> <= release out;
...
```

VHDL allows us to assign a composite value to a collection of signals by writing the collection in the form of an aggregate on the left-hand side of the assignment, for example:

```
(carry_out, sum) <= ('0' & a) + ('0' & b);
```

Note, in passing, that this form of aggregate assignment is legal in VHDL-2008 (see Section 6.4). We cannot, however, write an aggregate of signal names as the target of a force or release assignment to force or release each of the signal values. Instead, we must write a separate force or release assignment for each of the signals. For example, if we want to force and release the driving values of the two ports **carry_out** and **sum**, we would have to write:

```
sum <= force out unsigned'("00000000");
carry_out <= force out '1';
...

sum <= release out;
carry_out <= release out;
```

There is a further form of target signal for which we cannot write a force or release assignment. Suppose we define a resolved signal of a composite type, such as an array type. By that, we mean a signal with multiple sources, each of which is a composite value. The resolution function for the signal takes an array of composite values and determines a composite value as the resolved value of the signal. We cannot write a force or release assignment with an element of such a signal as the target. We can only force or release the signal as a whole. This mirrors the requirements that a process driving such a signal have a driver for all elements of the signal, and that sources for such a signal be sources for the entire signal. Note that resolved composite signals are different from signals of resolved elements, for example, signals of type **std_logic_vector**. We can force and release individual elements or slices of those signals, since each element is resolved individually.

Another case to consider is a force or release assignment written in a subprogram. VHDL has a rule that a signal assignment written in a procedure that is not contained within a process can only assign to a signal parameter of the procedure. The rationale is that assignment to a signal implies a driver for the signal. For signal parameters, the driver used is the driver for the actual signal provided by the process that calls the procedure. For other signals, a driver for the target signal would be implied for every process that calls the procedure. Identifying all of the callers in a large model would be very difficult. Moreover, if the procedure body is written separately from the calling processes, determining what drivers are created for a given process would be difficult. Thus, the restriction makes VHDL designs easier to analyze and understand. Force and release assignments, on the other hand, do not imply drivers. Rather, they would typically occur in testbench code, often referring to the target signals with external names. For these reasons, VHDL-2008 allows force and release assignments in procedures outside of processes to signals other than signal parameters.

One final aspect to discuss is the effect of multiple concurrent force and release assignments. Since they are sequential assignments written in processes, it is possible that multiple forces and releases could occur for a given signal during a single simulation cycle. The VHDL-2008 rules specify that if a force and release both occur, the effect is as though the release is immediately overridden by the force, and so the signal remains forced, but with the new force value. The effect of multiple forces is not defined. We should write our testbench models to avoid that occurring. The effect of multiple releases, however, is the same as a single release, and a release assignment on a signal that is not forced has no effect.

2.3 Context Declarations

Complex designs often call upon design units from several libraries and make use of several packages. As a consequence, each design unit in the design is preceded by a long list of library and use clauses, many of which are common to all of the design units. VHDL-2008 provides a new form of design unit, a *context declaration*, in which we can gather a collection of library and use clauses. We can refer to a context declaration before a design unit, rather than having to repeat the collection of library and use clauses. The form of a context declaration is

```
context identifier is
    ... -- library clauses, use clauses and context references
end context identifier;
```

Within a context declaration, we write library and use clauses in the same form as in a context clause preceding a design unit. We refer to a declared context with a context reference of the form

```
context context_name;
```

or, if we wish to refer to several context declarations:

```
context context_name, context_name, ...;
```

We can write a context reference in the context clause preceding a design unit, or nested within another context declaration. In each case, the context reference is equivalent to replacement by the list of library clauses and use clauses contained within the named context declaration.

EXAMPLE 2.6   *Organization-wide and project context declarations*

Suppose the methodology support team in Widgets, Inc., has assembled a library of reusable components, defined in a package **widget_comps** in a library with logical name **widget_lib**. The components package refers to a utility package, **widget_defs**, defining types and operations used across the organization. Both of these packages reference the standard **std_logic_1164** and **numeric_std** packages defined in library **IEEE**. The methodology team can provide a context declaration for use by projects in the organization:

```
context widget_context is
  library IEEE;
  use IEEE.std_logic_1164.all, IEEE.numeric_std.all;
  use widget_lib.widget_defs.all;
  use widget_lib.widget_comps.all;
end context widget_context;
```

This context declaration is analyzed into the **widget_lib** library. Given that a design needs to include a library clause for **widget_lib** in order to refer to the context declaration, there is no need to include that library clause in the context declaration itself. A design unit could reference the context declaration as follows:

```
library widget_lib;
context widget_lib.widget_context;
entity sample is
  ...
end entity sample;
```

Now suppose the Dongle project within Widgets, Inc., uses additional components provided by a third party, Gizmos Corp., defined by a package **gizmo_pkg** in library **gizmo_IP_lib**. The project also maintains a library **dongle_lib** for verified design units to be used in the project design flow, and a package **dongle_comps** with component declarations for the design units. The project's EDA support person can provide a context declaration for these libraries and packages, as well as referring to the organization's context declaration:

```
context dongle_context is
  library widget_lib;
  context widget_lib.widget_context;
  library gizmo_IP_lib;
  use gizmo_IP_lib.gizmo_pkg;
  use dongle_lib,dongle_comps.all;
end context dongle_context;
```

The EDA support person analyzes this context declaration into the **dongle_lib** library. A designer can then refer to the context in a design unit as follows:

```
library dongle_lib;
context dongle_lib.dongle_context;
entity frobber is
  . . .
end entity frobber;
```

The reference to **dongle_context** expands to include the reference to the organization's context and the library and use clauses for the third-party IP and the project repository. The reference to the organization's context in turn expands to include the library and use clauses for the standard packages and the organization's packages. Thus, the context clause written is equivalent to the following expanded context clause:

```
library dongle_lib;
library widget_lib;
library IEEE;
use IEEE.std_logic_1164.all, IEEE.numeric_std.all;
use widget_lib.widget_defs.all;
use widget_lib.widget_comps.all;
library gizmo_IP_lib;
use gizmo_IP_lib.gizmo_pkg;
use dongle_lib.dongle_comps.all;
entity frobber is
  . . .
end entity frobber;
```

VHDL uses library logical names to refer to physical design libraries. The mapping from a logical name to a physical library is implementation defined, and may vary between analysis of different design units. In order to avoid confusion when using context declarations, VHDL-2008 requires that a library logical name map to the same physical library during analysis of a context declaration and analysis of a reference to that context declaration. For example, if the logical name **gizmo_IP_lib** in Example 2.6 refers to **/home/dongle/gizmo/gizmo_IP_lib** when **dongle_context** is analyzed, the logical name must refer to the same physical library when entity **frobber** is analyzed.

As further reinforcement of this principle, we can't include a context clause before a context declaration, as we can for other design units. Thus, the following would be illegal:

```
library fizz_lib;  -- Illegal: precedes context declaration
context frazzle_ctx is
  use fizz_lib.fizz_pkg.all;
end context frazzle_ctx;
```

Instead, we should write the library clause inside the context declaration, so that it is included for any design unit that references the context declaration. Another related rule is that we cannot include a library clause referring to the working library, **WORK**, within a context declaration. Nor can we refer to the library name **WORK** in a use clause. The reason is that **WORK** is not defined for a context declaration, since context declarations don't have preceding context clauses.

Finally, VHDL-2008 defines two standard context declarations within the standard library **IEEE**:

```
context IEEE_BIT_CONTEXT is
  library IEEE;
  use IEEE.NUMERIC_BIT.all;
end context IEEE_BIT_CONTEXT;

context IEEE_STD_CONTEXT is
  library IEEE;
  use IEEE.STD_LOGIC_1164.all;
  use IEEE.NUMERIC_STD.all;
end context IEEE_STD_CONTEXT;
```

A design based on **bit** values might refer to the first of these context declarations, either in the context clause of a design unit or nested within a project context declaration. Similarly, a design based on **std_logic** values might refer to the second of these context declarations.

2.4 Integrated PSL

PSL is the IEEE Standard Property Specification Language (IEEE Std 1850). It allows specification of temporal properties of a model that can be verified either statically (using a formal proof tool) or dynamically (using simulation checkers). VHDL-2008 allows PSL code to be embedded as part of a VHDL model. This makes design for verification a much more natural activity, and simplifies development and maintenance of models. Since PSL is itself a significant language, we won't describe all of its features in detail in this book. Instead, we will just describe the way in which PSL can be embedded in VHDL. For a full description of PSL and its use in verifying designs, the interested reader is referred to other published books on the subject.[1]

---

1. See, for example, Cindy Eisner and Dana Fisman, *A Practical Introduction to PSL*, Springer, 2006.

In VHDL-2008 we can include PSL property, sequence, and default clock declarations in the declarative part of an entity, architecture, block, generate statement, or package declaration. We can then use the declared properties and sequences in PSL directives written in the statement parts of entities, architectures, blocks and generate statements.

Any properties that we write in PSL declarations and directives must conform to PSL's simple subset rules. In practice, this means that we can only write properties in which time moves forward from left to right through the property. Two examples from the PSL standard illustrate this. First, the following property is in the simple subset:

```
always (a -> next[3] b)
```

This property states that if **a** is true, then three cycles later, **b** is true; that is, time moves forward three cycles as we scan the property left to right. In contrast, the following property is not in the simple subset:

```
always ((a && next[3] b) -> c)
```

This property states that if **a** is true and **b** is true three cycles later, then **c** must have been true at the time **a** was true. The problem with this property is that time goes backward from **b** being true to **c** being true. A tool to check such a property is much more complex than one to check properties in the simple subset.

PSL directives require specification of a clock that determines when temporal expressions are evaluated. We can include a clock expression in a directive. However, since the same clock usually applies to all directives in a design, it is simpler to include a default clock declaration. If we write a default clock declaration in a region of a design, it applies to any PSL directives written in that region. We can include at most one default clock declaration in any given region.

There is one case where introduction of PSL embedded within VHDL leads to a possible ambiguity. Both PSL and VHDL include assert statements, but their meanings differ. If we write a statement of the form

```
assert not (a and b) report "a and b are both true";
```

it could be interpreted as a regular VHDL concurrent assertion statement that is to be checked whenever either of **a** or **b** changes value. Alternatively, in VHDL-2008, it could be interpreted as a PSL assert directive that requires the property **not** (a **and** b) to hold at time 0. In the interest of backward compatibility, VHDL-2008 interprets such ambiguous statements as regular VHDL concurrent assertion statements. If we really want to write a PSL assert directive of this form, we could modify the property so that it is unambiguously a PSL property, for example:

```
assert next[0] not (a and b) report "a and b are both true";
```

EXAMPLE 2.7   *Pipelined handshake assertion*

In their book *Assertion-Based Design*,[1] Foster *et al* describe a verification pattern for a system in which handshaking is pipelined. In their example, a system can receive up to 16 requests before acknowledging any of them. The system counts the number of requests and acknowledgments and includes an assertion that, for every request with a given request count, there is an acknowledgment with the same count within 100 clock cycles. We can describe the system in VHDL as follows:

```
library IEEE; context IEEE.IEEE_STD_CONTEXT;
entity slave is
  port ( clk, reset : in  std_logic;
         req        : in  std_logic;
         ack        : out std_logic;
         ... );
end entity slave;

architecture pipelined of slave is

  signal req_cnt, ack_cnt : unsigned(3 downto 0);

  default clock is rising_edge(clk);

  property all_requests_acked is
    forall C in {0 to 15}:
      always {req and req_cnt = C} |=>
             {[*0 to 99]; ack and ack_cnt = C};

begin

  req_ack_counter : process (clk) is
  begin
    if rising_edge(clk) then
      if reset = '1' then
        req_cnt <= "0000"; ack_cnt <= "0000";
      else
        if req = '1' then req_cnt <= req_cnt + 1; end if;
        if ack = '1' then ack_cnt <= ack_cnt + 1; end if;
      end if;
    end if;
  end process req_ack_counter;
  ...

  assert all_requests_acked;

end architecture pipelined;
```

The counters for requests and acknowledgments are implemented using the signals **req_cnt** and **ack_cnt** and the process **req_ack_counter**. We declare a property,

---

1. Harry D. Foster, Adam C. Krolnik, and David J. Lacey, *Assertion-Based Design*, Kluwer Academic Publishers, 2003.

all_requests_acked that expresses the verification condition for the design. We also include a default clock declaration for the architecture. It applies to the assert directive that we write in the statement part of the architecture, asserting that the verification condition holds.

In PSL, verification code can be written in verification units (**vunit**, **vprop** and **vmode** units) that are bound to instances of VHDL entities and architectures. VHDL-2008 considers such verification units as primary design units. Thus, they can be declared in VHDL design files and analyzed into VHDL design libraries.

A verification unit can include binding information that identifies a component instance to which directives apply. Alternatively, in VHDL-2008, we can bind a verification unit as part of the configuration of a design. One place to do that is in a configuration declaration. If we want to bind one or more verification units to the top-level entity in a configuration declaration, we include binding information as follows:

```
configuration config_name of entity_name is
    ...   -- use clauses, attribute specifications,
          -- group declarations
   use vunit verification_unit_name, ...;
   for architecture_name
      ...
   end for;
end configuration config_name;
```

Whenever the configuration declaration is instantiated, either at the top-level of a design hierarchy or as a component instance within a larger design, the named verification units are bound to the instance of the named entity and architecture. That means the names used in the verification units are interpreted in the context of the entity instance.

We can also bind verification units to component instances that are configured by a component configuration nested within a configuration declaration. The augmented form of component configuration, assuming the components are bound to an entity and architecture, and the architecture is further configured, is:

```
for instance_name, ... : component_name
  use entity entity_name(architecture_name);
  use vunit verification_unit_name, ...;
  for architecture_name
    ...
  end for;
end for;
```

In this case, the named verification units are bound to the instances specified in the component configuration.

The third place in which we can bind verification units in a VHDL design is in a configuration specification in the architecture or block where components are instantiated. The augmented form, again assuming components are bound to an entity and architecture, is:

```
for instance_name, ... : component_name
  use entity entity_name(architecture_name);
  use vunit verification_unit_name, ...;
end for;
```

This is similar to the form in a component configuration, but without the nested configuration for the architecture. Indeed, in order to make the syntax of a configuration specification more consistent with that of a component configuration, VHDL-2008 allows the reserved words **end for** to be used in a configuration specification even if there is no verification unit binding. On the other hand, if verification unit bindings are included, the **end for** reserved words are required.

Since a verification unit may include binding information as part of its declaration, there is potential for that information to conflict with binding information we write in a configuration. VHDL-2008 prevents such conflict by making it illegal to bind a verification unit in a configuration if the declaration of the unit already includes binding information. Hence, we would normally only write verification bindings in configurations for general-purpose verification units, and not for those written with particular instances in mind. In any case, it would be an error if we wrote a verification unit binding for a component instance that had no bound entity and architecture.

In addition to binding verification units directly in their declaration or indirectly in configurations, VHDL-2008 allows a tool to bind additional verification units through implementation-defined means. That might include command-line options, script commands, or selection using a graphical user interface.

EXAMPLE 2.8   *Binding a verification unit for complementary outputs*

Suppose we have a verification unit that ensures two outputs named Q and Q_n are complementary when sampled on rising edges of a signal named clk. The verification unit is:

```
vunit complementary_outputs {
  assert always Q = not Q_n;
}
```

We can bind this verification unit to various parts of a design. First, a gate-level model of a D Flip-flop might be described as follows:

```
entity D_FF is
  port ( clk, reset, D : in  bit;
         Q, Q_n        : out bit );
end entity D_FF;

architecture gate_level of D_FF is
  component and2 is ...
  ...
begin
  G1 : and2 ...
```

```
    . . .
end architecture gate_level;
```

A configuration declaration for the D flip-flop can bind the verification unit to the top-level entity as follows:

```
configuration fast_sim of D_FF is
  use vunit complementary_outputs;
  for gate_level
    for all : and2
      . . .
    end for;
    . . .
  end for;
end configuration fast_sim;
```

We could then instantiate the configuration in a design, and for each instance, the verification unit **complementary_outputs** would be bound to the instantiated entity and architecture.

Second, suppose we instantiate a parallel-in/serial-out shift register within an RTL design:

```
entity system is
  . . .
end entity system;

architecture RTL of system is
  component shift_reg is
    . . .
  end component shift_reg;
  . . .
begin
  serializer : shift_reg ...;
  . . .
end architecture RTL;
```

We can write a configuration declaration that binds an entity and architecture to the component instance and that also binds the **complementary_outputs** verification unit:

```
configuration verifying of system is
  for RTL
    for serializer : shift_reg
      use entity work.shift_reg(RTL);
      use vunit complementary_outputs;
    end for;
  end for;
end configuration verifying;
```

Third, we could specify the same binding information directly in the architecture, rather than in a separate configuration. The revised architecture is:

```
architecture RTL of system is
  component shift_reg is

    ...

  end component shift_reg;
  for serializer : shift_reg
    use entity work.shift_reg(RTL);
    use vunit complementary_outputs;
  end for;
begin
  serializer : shift_reg ...;

  ...

end architecture RTL;
```

There are some further points to make about PSL embedded in VHDL. First, since we can declare properties and sequences within VHDL, we can also specify attribute values for them. To that end, we can use the reserved words **property** and **sequence** in attribute specifications for declared properties and sequences, respectively. For example:

```
property SingleCycleRequest is
  always req -> next not req;
sequence ReadCycle is
  { ba; {bb[*]} && {ar[->]; dr[->]}; not bb };

attribute enable_heuristics of
          SingleCycleRequest : propery is true;
attribute enable_heuristics of ReadCycle : sequence is true;
```

Second, PSL has a rich set of reserved words, some of which may conflict with VHDL identifiers. In VHDL-2008, the following PSL keywords are VHDL reserved words, and cannot be used as identifiers:

```
assert
assume
assume_guarantee
cover
default
fairness
property
restrict
restrict_guarantee
sequence
strong
vmode
```

```
vprop
vunit
```

Other PSL reserved words are only recognized as such within VHDL code when they occur in PSL declarations and directives. They can be used as VHDL identifiers, but such identifiers are hidden within PSL declarations and directives. For example, we can legally write the following declaration:

function rose (x : boolean) **return** boolean **is** ...;

But if we then declare a sequence:

sequence cover_fifo_empty **is**
  {reset_n && **rose**(cnt = 0)};

The reference to **rose** in the sequence declaration is to the PSL built-in function, not to the declaration written in VHDL.

Finally, PSL includes features for declaring and instantiating macros, and allows for preprocessor directives. These features can only be used in PSL verification units, not in other VHDL design units.

2.5   IP Encryption

As designs become more complex, designers are increasingly using intellectual property (IP) provided by IP vendors. IP providers invest considerable effort in developing their products, and may be loath to release them without protecting their investment. From the IP provider's point of view, there are two potential places where their IP may be compromised. First, the IP provider must transmit the IP to a customer. During that process, a malicious third party could eavesdrop on the transmission and intercept the IP. Second, the customer must receive, store, and use the IP. During that process, an unscrupulous customer could reuse the IP without compensating the provider. Hence, the customer is technically treated as a malicious third party, though it would not be politic to express the relationship in those terms! The real recipient of the IP is the customer's tool, which must use the IP only in ways approved by the IP provider and must avoid disclosing the IP to the customer.

One way of protecting IP is for the provider to encrypt it in a form that can be decrypted and used by a customer's tools, but that cannot be read by the customer. VHDL-2008 provides a flexible set of features to support such protection. Before we describe them in detail, we will first review some of the basic principles and protocols for encryption so that we can understand how to use the language features.

Information to be communicated between two parties can be protected by transforming it with a *cipher*. A cipher is a function that takes *plain text* and a string of bits called a *key* as input and produces *cipher text* as output. This process is called *encryption*. The reverse process, *decryption*, takes the cipher text and a key as input and reproduces the original plain text. The quality of a cipher is determined by the difficulty of determining the plain text from the cipher text without the key. A good cipher will yield significantly different cipher text for minor changes to the key used for encryption.

There are two forms of cipher in widespread use. A *symmetric cipher* uses the same key for both encryption and decryption. The key is called a *secret key*, since it must be kept secret between the communicating parties. Should the secret be revealed to a third party, they could decrypt any intercepted encrypted information. Examples of symmetric ciphers are the Data Encryption Standard (DES), and the Advanced Encryption Standard (AES).

An *asymmetric cipher* uses a pair of related keys, one for encryption and the other for decryption. Key pairs are generated in such a way that it is infeasible to determine either key from the other. Information encrypted with one key of a pair can only be decrypted with the other key of the pair. Examples of asymmetric ciphers are RSA and ElGamal. Asymmetric ciphers are used in protocols where each communicating party generates a key pair. They keep one key of the pair, the *private key*, secret. They publish the other key, the *public key*, through some means of dissemination that associates the public key with the communicating party's identity. For example, they might publish it on their web site. A sender of information can use an asymmetric cipher to protect information destined for a recipient. The sender encrypts the information using the recipient's public key. Only the recipient can then decrypt the information, since only they have the corresponding private key.

While asymmetric ciphers can yield more secure communication, they involve significantly greater computation than symmetric ciphers. For that reason, most applications involving asymmetric ciphers use a two-stage encryption process called a *digital envelope*. First a *session key* is randomly generated, for use in one communication session only. Next, that session key is used with a symmetric cipher to encrypt the information. In order to communicate the session key to the recipient so that they can decrypt the information, the session key is encrypted using an asymmetric cipher with the recipient's public key, and sent to the recipient. They are the only party able to decrypt the session key, since only they have the right private key. They can then proceed to decrypt the communicated information using the symmetric cipher with the decrypted session key. The advantage of this approach is that only a relatively small amount of information (the session key) need be processed using the computationally intensive asymmetric cipher. The bulk of the information is processed using the lighter-weight symmetric cipher.

One problem that arises in protected communication is the need to verify that received information did in fact originate with the purported sender, and that it was not changed in transit (either by corruption or maliciously) by a third party. This problem is addressed by having the sender transmit a *digital signature* for the information. The sender uses a *hash function* to compute a *digest* of the information. A hash function takes a (potentially large) string of bit as input and produces a small string of bits, the digest, that depends on all of the input bits. A good hash function has the property that two distinct input strings are highly unlikely to yield the same output string. Examples of hash functions include SHA1, MD2, MD5, and RIPEMD. Having computed the digest of the information, the sender encrypts it using an asymmetric cipher with their private key and transmits the result as the digital signature. A recipient decrypts the signature using the purported sender's public key to retrieve the digest. The recipient also independently calculates the digest of the received information using the hash function. If the two digests are the same, the information has been received correctly, since only the real sender's public key could decrypt the digest correctly, and only the real information would yield the same digest. If, on the other hand, the digests differ, then either the transmitted digest was encrypted with someone else's key, or the transmitted message was changed. Either way, the transmission was compromised, and the recipient knows not to trust the received information.

If we are to apply cryptographic techniques to transmission of VHDL models, we need to consider the way in which the encrypted information is encoded. Plain-text VHDL models consist of printable ASCII or Latin-1 characters and are immune to the way ends of lines are encoded. Consequently, we can store and transmit plain-text models through a variety of media without being concerned about encodings. However, the process of encryption produces a string of bits, which cannot be guaranteed to be interpreted as printable characters. We cannot reliably transmit the encrypted model, since some media might transform sequences of bits interpreted as line ends, or might interpret sequences of bits as in-band control codes. To avoid these problems, we can encode the encrypted model using an encoding method that uses printable characters to represent the string of bits. A sender encrypts information and encodes it for transmission, and a recipient decodes the received information and decrypts the result. Examples of encoding methods include uuencode, base64, and quoted-printable, all of which are described by Internet message-transfer standards.

With this overview of cryptography in hand, we can now discuss the features provided in VHDL-2008 to support cryptographic protection of IP. The features use a standard set of tool directives (see Section 9.21). A tool directive is an annotation included in a VHDL design file that provides information to a tool processing the VHDL design. It does not logically form part of the design itself. For IP protection, VHDL-2008 defines *protect directives* that are used by an IP provider's *encryption tool* to govern encryption of sections of a VHDL design and by a customer's *decryption tool* to decrypt those sections. The decryption tool is typically a simulator, synthesis tool, or some other tool that deals with VHDL code. It uses the decrypted sections of the design, but does not store them in any form that could be revealed to the customer. Protect directives each takes one of three forms:

```
`protect keyword
`protect keyword = value
`protect keyword = ( keyword = value, ... )
```

Like any tool directive, a protect directive starts with the "tick" symbol, and ends at the end of the line. The keyword or keywords in a protect directive identify the kind of information conveyed by the directive. Note that we write the keywords in boldface here to indicate that they have special meanings in protect directives. They are not reserved words outside of protect directives. The values are literal expressions of various types. If we have a number of consecutive protect directives, we can merge them into a single directive. Thus, we can write the sequence of directives

```
`protect keyword1 = value1
`protect keyword2 = value2
`protect keyword3
```

equivalently as

```
`protect keyword1 = value1, keyword2 = value2, keyword3
```

An IP provider starts the process by identifying one or more sections of a VHDL design file that they want to protect. They edit the design file to wrap each section in an *encryption envelope*, consisting of one or more protect directives at the start of the section, and a closing protect directive at the end of the section. The simplest form of encryption envelope is:

```
`protect begin
... -- protected source code in plain-text form
`protect end
```

This simply delimits the protected source code, and assumes an encryption tool will use default information about the ciphers, keys and encoding for encryption. More elaborate encryption envelopes precede the **begin** directive with protect directives specifying ciphers, keys, encoding and other optional information.

The IP provider then processes the design file with an encryption tool to produce a version of the design file with each encryption envelope replaced by a corresponding *decryption envelope* of the following form:

```
`protect begin_protected
protect directives and encoded encrypted information
`protect end_protected
```

We will use a series of examples to show how the various directives are used to form encryption and decryption envelopes for various use cases. In each case we will assume that the decryption tool has access to the required keys, and that the encryption tool knows about those keys. We will return to the topic of key exchange in Section 2.5.1.

EXAMPLE 2.9 *Simple encryption envelope with symmetric cipher*

In one of the simplest use cases, an IP provider wants to provide protected IP to a customer for use with a single tool. We can use a symmetric cipher, for which the key is known to both the IP provider and to the customer's decryption tool. The IP provider wraps the protected section in the source code in an encryption envelope, as follows:

```
entity accelerator is
  port ( ... );
end entity accelerator;

architecture RTL of accelerator is
`protect data_keyowner = "ACME IP User"
`protect data_keyname  = "ACME Sim Key"
`protect data_method   = "aes192-cbc"
`protect encoding      = (enctype = "base64")
`protect begin
  signal ...
begin
  process ...
```

```
   ...
  `protect end
  end architecture RTL;
```

The IP provider leaves the information about the entity's interface and the name of the architecture unprotected so that the customer can instantiate the design. The entire inner workings of the architecture, however, are not to be revealed to the customer. The **data_keyowner** and **data_keyname** directives specify identifiers that the encryption and decryption tools can use to retrieve the key. The **data_method** directive specifies the cipher to use for encryption and decryption, and the **encoding** directive specifies the method to use to encode the cipher text produced by the encryption tool.

The IP provider processes the original source code file with an encryption tool, which produces a modified file with the encryption envelope replaced by a decryption envelope:

```
entity accelerator is
  port ( ... );
end entity accelerator;

architecture RTL of accelerator is
`protect begin_protected
`protect encrypt_agent       = "Encryptomatic"
`protect encrypt_agent_info = "2.3.4a"
`protect data_keyowner = "ACME IP User"
`protect data_keyname  = "ACME Sim Key"
`protect data_method   = "aes192-cbc"
`protect encoding=(enctype="base64", line_length=40, bytes=4006)
`protect data_block
encoded cipher-text
...
`protect end_protected
end architecture RTL;
```

The **encrypt_agent** and **encrypt_agent_info** directives provide information about the encryption tool. This can help in tracking down any problems that might arise. The directives specifying the key, cipher, and encoding method are replicated in the decryption envelope. In the case of the **encoding** directive, further information about the maximum line length for the encoded cipher text and the number of bytes of cipher text is also provided. The encoded cipher text then starts immediately after the **data_block** directive. The **end_protected** directive marks the end of the decryption envelope.

EXAMPLE 2.10 *Digital envelope encrypted for a single customer*

One of the problems with using a symmetric cipher to encrypt IP is that the risk of the secret key being divulged. We can avoid that risk by using a digital envelope to transmit the IP. The IP provider includes directives in the encryption envelope to specify a cipher and key to use to encrypt a session key. The IP provider can also specify the symmetric cipher to use to encrypt the data with the session key. The design file with the revised encryption envelope is:

```
entity accelerator is
  port ( ... );
end entity accelerator;

architecture RTL of accelerator is
`protect key_keyowner = "ACME IP User"
`protect key_keyname  = "ACME Sim Key"
`protect key_method   = "rsa"
`protect key_block
`protect data_method  = "aes192-cbc"
`protect encoding     = (enctype = "base64")
`protect begin
  signal ...
begin
  process ...
  ...
`protect end
end architecture RTL;
```

The **key_keyowner** and **key_keyname** directives specify identifiers that the encryption tool can use to retrieve the customer's public key. The **key_method** directive specifies the cipher to use to encrypt the session key. The **key_block** directive marks the end of the key information. Its presence in the encryption envelope specifies use of a digital envelope, since the preceding key directives can be omitted, implying default values. The **data_method** directive specifies the cipher to use for encryption and decryption with the session key. The **encoding** directive specifies the method to use to encode both the encrypted session key and the encrypted section of the model.

The IP provider processes this source code file with an encryption tool, which generates a session key and produces a modified file with the encryption envelope replaced by a decryption envelope specifying a digital envelope:

```
entity accelerator is
  port ( ... );
end entity accelerator;

architecture RTL of accelerator is
`protect begin_protected
```

```
`protect encrypt_agent      = "Encryptomatic"
`protect encrypt_agent_info = "2.3.4a"
`protect key_keyowner = "ACME IP User"
`protect key_keyname  = "ACME Sim Key"
`protect key_method   = "rsa"
`protect encoding=(enctype="base64", line_length=40, bytes=256)
`protect key_block
encoded cipher-text for session key
`protect data_method  = "aes192-cbc"
`protect encoding=(enctype="base64", line_length=40, bytes=4006)
`protect data_block
encoded cipher-text for model code
...
`protect end_protected
end architecture RTL;
```

The directives specifying the key and cipher for encrypting the session key are replicated in the decryption envelope. The **encoding** directive is also replicated to specify the encoding for the encrypted session key, augmented with information about the maximum line length for the encoded cipher text and the number of bytes in the encrypted session key. The encoded cipher text for the session key then starts immediately after the **key_block** directive. Next, the **data_method** directive specifying the cipher for the model code is replicated in the decryption envelope. The **encoding** directive is also replicated here, augmented with information about the maximum line length and the number of bytes. The encoded cipher text for the model code starts immediately after the **data_block** directive. The **end_protected** directive marks the end of the decryption envelope.

EXAMPLE 2.11   *Digital envelope encrypted for multiple customers or tools*

In Example 2.9 and Example 2.10, the IP is encrypted in a form that can be decrypted by a single customer or by a single tool. If the IP provider wants to distribute the IP to multiple customers or to a customer for use with multiple tools, he or she would have to generate multiple versions using the encryption tool, once per customer. We can avoid this repetition by using a variation on the digital envelope approach. Again, we specify that a session key be used to encrypt the model code. However, that session key is then encrypted multiple times, once per customer or customer's tool. The revised source file with the encryption envelope is:

```
entity accelerator is
  port ( ... );
end entity accelerator;

architecture RTL of accelerator is
`protect key_keyowner = "ACME IP User1"
`protect key_keyname  = "ACME Sim Key"
```

```
`protect key_method    = "rsa"
`protect key_block
`protect key_keyowner = "ACME IP User2"
`protect key_keyname  = "ACME Synth Key"
`protect key_method    = "elgamal"
`protect key_block
`protect key_keyowner = "ACME IP User3"
`protect key_keyname  = "ACME P&R Key"
`protect key_method    = "aes192-cbc"
`protect key_block
`protect data_method   = "aes192-cbc"
`protect encoding      = (enctype = "base64")
`protect begin
  signal ...
begin
  process ...
  ...
`protect end
end architecture RTL;
```

Each group of key directives specifies information for encryption of a session key for decryption by a given decryption tool. The first two groups specify encryption using asymmetric ciphers, as is normally done in digital envelopes. However, we can also use a symmetric cipher to encrypt the session key, as specified in the third group of key directives.

As in the earlier examples, the IP provider processes this source code file with an encryption tool, which generates a session key and produces a modified file with the encryption envelope replaced by a decryption envelope specifying a digital envelope:

```
entity accelerator is
  port ( ... );
end entity accelerator;

architecture RTL of accelerator is
`protect begin_protected
`protect encrypt_agent      = "Encryptomatic"
`protect encrypt_agent_info = "2.3.4a"
`protect key_keyowner = "ACME IP User1"
`protect key_keyname  = "ACME Sim Key"
`protect key_method    = "rsa"
`protect encoding=(enctype="base64", line_length=40, bytes=256)
`protect key_block
encoded cipher-text for session key
`protect key_keyowner = "ACME IP User2"
`protect key_keyname  = "ACME Synth Key"
`protect key_method    = "elgamal"
```

```
`protect encoding=(enctype="base64", line_length=40, bytes=256)
`protect key_block
encoded cipher-text for session key
`protect key_keyowner = "ACME IP User3"
`protect key_keyname  = "ACME P&R Key"
`protect key_method   = "aes192-cbc"
`protect encoding=(enctype="base64", line_length=40, bytes=256)
`protect key_block
encoded cipher-text for session key
`protect data_method  = "aes192-cbc"
`protect encoding=(enctype="base64", line_length=40, bytes=4006)
`protect data_block
encoded cipher-text for model code
...
`protect end_protected
end architecture RTL;
```

In this case, the decryption envelope includes a group of key directives and a key block corresponding to each group of key directives in the encryption envelope. Each of the targeted decryption tools, when it processes the decryption envelope, checks whether it has access to the key specified by each group of key directives. If it has one of the keys, it can use that key to decrypt the session key, and thus decrypt the model code.

EXAMPLE 2.12 *Digital signature for authentication of the provider*

Suppose our IP provider delivers encrypted IP by making it available for download from a file server. They use our public key to deliver the IP in digital envelope form. An unscrupulous third-party IP provider could seek to besmirch the name of our trusted IP provider by spoofing their server and providing a buggy version of the IP. Since the IP is encrypted using our public key, which is widely known, we would not be aware of the switch.

The solution is for our trusted IP provider to include a digital signature in the delivered model. The encryption envelope, revised from that in Example 2.11, is:

```
entity accelerator is
  port ( ... );
end entity accelerator;

architecture RTL of accelerator is
`protect key_keyowner = "ACME IP User"
`protect key_keyname  = "ACME Sim Key"
`protect key_method   = "rsa"
`protect key_block
`protect data_method  = "aes192-cbc"
`protect digest_keyowner  = "GoodGuys IP Author"
```

```
`protect digest_keyname     = "GoodGuys Signing Key"
`protect digest_key_method = "rsa"
`protect digest_method      = "sha1"
`protect digest_block
`protect encoding = (enctype = "base64")
`protect begin
  signal ...
begin
  process ...
  ...
`protect end
end architecture RTL;
```

The digest directives in the encryption envelope specify that a digital signature should be generated for the model code contained in the envelope. The **digest_method** directive specifies the hash function for computing the digest, and the **digest_keyowner**, **digest_keyname** and **digest_key_method** directives specify the cipher and key to use to encrypt the digest. The **digest_key_method** directive must specify an asymmetric cipher, since digital signatures are predicated on the use of such ciphers.

The IP provider processes this source code file with an encryption tool, which computes and encrypts the digest to form the digital signature. It uses the private key of the key pair specified by the digest key directives. It includes the digest in the decryption envelope corresponding to the encryption envelope:

```
entity accelerator is
  port ( ... );
end entity accelerator;

architecture RTL of accelerator is
`protect begin_protected
`protect encrypt_agent       = "Encryptomatic"
`protect encrypt_agent_info = "2.3.4a"
`protect key_keyowner = "ACME IP User"
`protect key_keyname  = "ACME Sim Key"
`protect key_method   = "rsa"
`protect encoding=(enctype="base64", line_length=40, bytes=256)
`protect key_block
encoded cipher-text for session key
`protect data_method   = "aes192-cbc"
`protect encoding=(enctype="base64", line_length=40, bytes=4006)
`protect data_block
encoded cipher-text for model code
...
`protect digest_keyowner   = "GoodGuys IP Author"
`protect digest_keyname    = "GoodGuys Signing Key"
`protect digest_key_method = "rsa"
```

```
`protect digest_method     = "sha1"
`protect digest_block
`protect encoding=(enctype="base64", line_length=40, bytes=16)
`protect digest_block
encoded cipher-text for digest
...
`protect end_protected
end architecture RTL;
```

Our trusted IP provider places this model on the file server for us to download. Now suppose the unscrupulous third-party IP provider performs their network hack and substitutes a buggy model. In their first attempt, they substitute the buggy code, encrypted with a session key that they generate, and encrypt the session key with our public key. Our decryption tool successfully decrypts the session key and uses it to decrypt the model. However, since we want to verify that we have the right model, the decryption tool computes the digest of the decrypted model using the hash function specified in the **digest_method** directive. The tool also uses the public key of the key pair identified in the digest key directives to decrypt the transmitted digest. Since the model code is different from the original code provided by the trusted IP provider, the two digests are not the same. Our decryption tool alerts us to this fact, and we contact our trusted IP provider to attempt to remedy the problem.

Now suppose the unscrupulous third-party IP provider realizes their ruse was unsuccessful, and tries a different tack. As well as substituting the buggy model, suitably encrypted, they also generate a digital signature for the buggy model and substitute it for the real digital signature. They use their own private key to encrypt the digest, and include digest key directives that identify their key pair. Again, our decryption tool successfully decrypts the model and calculates the digest. The tool also attempts to decrypt the transmitted digest in order to compare with the computed digest. At this point, there are two possible outcomes. First, if the tool does not have access to the unscrupulous provider's public key, it will be unable to proceed and will warn us that it could not verify the digital signature. Alternatively, if the tool does have access to the unscrupulous provider's public key, it will use it to decrypt the transmitted digest and compare it with the computed digest. In this case, the digests will match. It will be up to us to check that signature verification was performed with the key we expected. This illustrates that we need to be vigilant when checking digital signatures, so that we are not duped by a simple key substitution. We will discuss this more in Section 2.5.1, where we address the issue of key exchange.

EXAMPLE 2.13   *Viewport for accessing a declaration in a protected model*

An IP provider may wish to allow limited access to some items declared within the protected source code. In Examples 2.1 and 2.2 in Section 2.1, we showed a testbench monitoring the internal state of the control section of a design under verification. An IP provider can allow such access by including a **viewport** directive in the encryption envelope. An example is:

```
entity accelerator is
  port ( ... );
end entity accelerator;

architecture RTL of accelerator is
`protect data_keyowner = "ACME IP User"
`protect data_keyname  = "ACME Sim Key"
`protect data_method   = "aes192-cbc"
`protect encoding       = (enctype = "base64")
`protect viewport=(object="accelerator:RTL.state", access="RW");
`protect begin
  signal state : std_logic_vector(3 downto 0);
  ...
begin
  process ...
  ...
`protect end
end architecture RTL;
```

While most of the inner workings of the architecture are not revealed to the customer, the **viewport** directive provides the pathname of the object representing the
internal state signal and grants read/write access. The IP provider processes the
source code file with an encryption tool, which includes the **viewport** directive in
the decryption envelope:

```
entity accelerator is
  port ( ... );
end entity accelerator;

architecture RTL of accelerator is
`protect begin_protected
`protect encrypt_agent       = "Encryptomatic"
`protect encrypt_agent_info = "2.3.4a"
`protect viewport=(object="accelerator:RTL.state", access="RW");
`protect data_keyowner = "ACME IP User"
`protect data_keyname  = "ACME Sim Key"
`protect data_method   = "aes192-cbc"
`protect encoding=(enctype="base64", line_length=40, bytes=4006)
`protect data_block
encoded cipher-text
...
`protect end_protected
end architecture RTL;
```

The customer can instantiate the IP in a design and use an external name to refer
to the state signal:

```vhdl
architecture monitoring of tb is
  ...
begin
  ... -- clock and reset generation

  accelerator_duv : entity work.accelerator(rtl)
    port map ( ... );

  monitor : process (clk) is
    use std.textio.all;
    file state_file : text open write_mode is state_file_name;
    alias accelerator_state is
      <<signal accelerator_duv.state :
                std_logic_vector(3 downto 0)>>;
  begin
    if falling_edge(clk) then
      write(L, accelerator_state); writeline(state_file, L);
    end if;
  end process monitor;

end architecture monitoring;
```

While the **viewport** directive provides access to the internal signal, it does not provide complete information. The IP provider would also need to provide documentation describing the type of the signal and other relevant information.

Now that we have seen how protection envelopes are formed in various scenarios, we will describe the details of VHDL-2008's IP protection mechanism. As we have mentioned, it is based on a set of tool directives. The full list of directives is as follows:

`protect begin

Indicates the beginning of the source code to be encrypted in an encryption envelope.

`protect end

Indicates the end of an encryption envelope.

`protect begin_protected

Indicates the beginning of a decryption envelope.

`protect end_protected

Indicates the end of a decryption envelope.

`protect author = "*author name*"

> Identifies the author of the protected IP. If this directive appears in an encryption envelope, the encryption tool copies it unchanged to the corresponding decryption envelope.

`protect author_info = "*author info*"

> Provides further information about the author of the protected IP, such as an organization name or contact details. If this directive appears in an encryption envelope, the encryption tool copies it unchanged to the corresponding decryption envelope.

`protect encrypt_agent = "*encrypt agent name*"

> This directive must appear in a decryption envelope, and identifies the encryption tool that produced the decryption envelope.

`protect encrypt_agent_info = "*encrypt agent info*"

> This directive may appear in a decryption envelope, and provides further information about the encryption tool that produced the decryption envelope.

`protect key_keyowner = "*key owner name*"

> Identifies the owner of a key or key pair used to encrypt a session key.

`protect key_keyname = "*key name*"

> Used together with the key owner name to identify a particular key or key pair used to encrypt a session key.

`protect key_method = "*cipher name*"

> Specifies the cipher used to encrypt a session key.

`protect key_block

> In an encryption envelope, specifies use of a digital envelope. In the corresponding decryption envelope, indicates the beginning of the encoded cipher text of the session key.

`protect data_keyowner = "*key owner name*"

> Identifies the owner of a key or key pair used to encrypt the source code.

`protect data_keyname = "*key name*"

> Used together with the key owner name to identify a particular key or key pair used to encrypt the source code.

`protect data_method = "*cipher name*"

> Specifies the cipher used to encrypt the source code.

`protect data_block

In a decryption envelope, indicates the beginning of the encoded cipher text of the source code.

`protect digest_keyowner = "*key owner name*"

Identifies the owner of the key pair used to encrypt the digest in a digital signature.

`protect digest_keyname = "*key name*"

Used together with the key owner name to identify a particular key pair used to encrypt the digest in a digital signature.

`protect digest_key_method = "*cipher name*"

Specifies the asymmetric cipher used to encrypt the digest in a digital signature.

`protect digest_method = "*hash function name*"

Specifies the hash function used to compute the digest in a digital signature.

`protect digest_block

In an encryption envelope, specifies use of a digital signature. In the corresponding decryption envelope, indicates the beginning of the encoded cipher text of the digest.

`protect encoding =
(**enctype** = "*encoding name*", **line_length** = *integer*, **bytes** = *integer*)

In an encryption envelope, this directive specifies the encoding to be used for cipher text in the corresponding decryption envelope. The **line_length** keyword and value are optional and specify the maximum line length for encoded text. Text longer than this amount is split into multiple lines. The **bytes** keyword and value are also optional and are ignored in an encryption envelope in any case.

In a decryption envelope, this directive appears preceding each key, data, and digest block. It specifies the encoding, maximum line length, and number of bytes of cipher text encoded in the block.

`protect viewport = (**object** = "*object pathname*", **access** = "*access type*")

Identifies an object declared within the protected source code for which access is granted. If this directive appears in an encryption envelope, the encryption tool copies it unchanged to the corresponding decryption envelope.

The pathname consists of the names of regions enclosing the declaration, starting with the design unit name and continuing with the names of nested regions, separated by "." characters, for example,

```
"my_entity.cycle_monitor.cycle_count"
```

If the object is declared within an architecture, the design unit name is the combination of the entity name and the architecture name, separated by a colon, for example,

```
"my_entity:RTL.current_state"
```

If the object is declared within a package body, the design unit name consists of the package name, followed by ":**body**", for example,

```
"IP_pkg:body.trace_file"
```

The access type string must be one of "R", "W", or "RW" (or the lowercase equivalents), indicating read access, write access, or read/write access, respectively.

`` `protect decrypt_license = ``
(**library** = "*library name*",
 entry = "*acquisition routine name*", **feature** = "*feature name*",
 exit = "*release routine name*", **match** = *integer*)

This directive specifies information for acquiring a decryption license. If the directive appears in an encryption envelope, the encryption tool copies it unchanged to the corresponding decryption envelope. If the directive appears in a decryption envelope, a decryption tool must attempt to acquire the specified license. If acquisition is successful, it continues decrypting the model. Otherwise, it is expected to stop further decryption.

The library name string identifies the object library in the decryption tool's host file system containing routines for license management. The tool should call the routine identified by the acquisition routine name, passing the feature name string as an argument, to acquire a license. The tool should compare the return value of the routine with the match integer. If they are equal, acquisition succeeded. When the tool has completed decryption, it should relinquish the license by calling the routine identified by the release routine name.

`` `protect runtime_license = ``
(**library** = "*library name*",
 entry = "*acquisition routine name*", **feature** = "*feature name*",
 exit = "*release routine name*", **match** = *integer*)

This directive specifies information for acquiring a runtime license. If the directive appears in an encryption envelope, the encryption tool copies it unchanged to the corresponding decryption envelope. If the directive appears in a decryption envelope, a decryption tool must attempt to acquire the specified license. If acquisition is successful, the tool may continue with analysis and execution of the model. Otherwise, it is expected not to execute the model. The information in this directive is the same as that in a **decrypt_license** directive.

`**protect comment** = "*comment string*"`

> This directive allows the IP author to provide comments in the model. If the directive appears in an encryption envelope, either preceding or within the source code, the encryption tool copies it unchanged to the corresponding decryption envelope. If it is within the source code, the encryption tool skips over it when encrypting the source code.

Several directives use strings to specify ciphers, encodings, and hash functions. VHDL-2008 defines particular string values for these directives. If a tool supports the given cipher, encoding, or hash function, it must use the defined string value to specify it. A tool may also support other methods, in which case it uses an implementation-defined string value. Table 2.1 shows the string values for specifying ciphers. Every tool must support at least the DES cipher. Table 2.2 shows the string values for specifying encodings. Every tool must support at least uuencode and base64. Table 2.3 shows the string values for specifying hash functions. Every tool must support at least SHA1 and MD5.

TABLE 2.1   *Strings for specifying ciphers*

String	Cipher	Cipher type
"des-cbc"	DES in CBC mode.	Symmetric
"3des-cbc"	Triple DES in CBC mode.	Symmetric
"aes128-cbc"	AES in CBC mode with 128-bit key.	Symmetric
"aes192-cbc"	AES in CBC mode with 192-bit key.	Symmetric
"aes256-cbc"	AES in CBC mode with 256-bit key.	Symmetric
"blowfish-cbc"	Blowfish in CBC mode.	Symmetric
"twofish128-cbc"	Twofish in CBC mode with 128-bit key.	Symmetric
"twofish192-cbc"	Twofish in CBC mode with 192-bit key.	Symmetric
"twofish256-cbc"	Twofish in CBC mode with 256-bit key.	Symmetric
"serpent128-cbc"	Serpent in CBC mode with 128-bit key.	Symmetric
"serpent192-cbc"	Serpent in CBC mode with 192-bit key.	Symmetric
"serpent256-cbc"	Serpent in CBC mode with 256-bit key.	Symmetric
"cast128-cbc"	CAST-128 in CBC mode.	Symmetric
"rsa"	RSA.	Asymmetric
"elgamal"	ElGamal.	Asymmetric
"pgp-rsa"	OpenPGP RSA key.	Asymmetric

TABLE 2.2   *Strings for specifying encodings*

String	Encoding methods
"uuencode"	IEEE Std 1003.1™-2001 (uuencode Historical Algorithm)
"base64"	IETF RFC 2045, also IEEE Std 1003.1 (uuencode -m)
"quoted-printable"	IETF RFC 2045
"raw"	Identity transformation; no encoding is performed, and the data may contain non-printing characters.

TABLE 2.3   *Strings for specifying hash functions*

Digest method string	Required/optional	Hash function
"sha1"	Required	Secure Hash Algorithm 1 (SHA-1).
"md5"	Required	Message Digest Algorithm 5.
"md2"	Optional	Message Digest Algorithm 2.
"ripemd-160"	Optional	RIPEMD-160.

We can now describe the rules for forming an encryption envelope in a model. The rules allow for considerable flexibility, but we must at least include the **begin** and **end** directives to mark out the source code to be encrypted.

We can precede the **begin** directive with a **key_block** directive if we want to specify use of digital envelopes. We can specify the cipher and key to use to encrypt the session key by including a **key_method** and a **key_keyowner** directive (and optionally a **key_keyname** directive). If we don't specify the cipher and key, the encryption tool chooses a default cipher and key. The **key_method**, **key_keyowner** and **key_keyname** directives can appear in any order, but must immediately precede the **key_block** directive. We can include more than one group of key-related directives, as we described in Example 2.11.

We can also precede the **begin** directive with a **data_method** directive if we want to specify the cipher to use to encrypt the source code. If we are not using digital envelopes and we include a **data_method** directive, we must also include a **data_keyowner** directive and optionally a **data_keyname** directive to identify the key. If we are using digital envelopes, the encryption tool generates the session key, so we do not include directives to identify the key. If we omit the **data_method** directive, the encryption tool chooses a default cipher. All of the directives relating to encryption of the source code must appear together in an encryption envelope.

If we want to include a digital signature, we precede the **begin** directive with a **digest_block** directive. We can specify the cipher and key to use to encrypt the digest by including a **digest_key_method** and a **digest_keyowner** directive (and optionally a **digest_keyname** directive). If we don't specify the cipher and key, the encryption tool chooses a default cipher and key. Similarly, we can specify the hash function to use by including a **digest_method** directive. If we don't specify a hash function, the encryption tool chooses a default hash function. The **digest_key_method**, **digest_keyowner**, **digest_keyname**, and **digest_method** directives can appear in any order, but must immediately precede the **digest_block** directive.

Beyond these specifications, we can include directives to identify the IP author, describe licenses and viewports, and specify the encoding to use. If we don't specify the encoding, the encryption tool chooses a default encoding. We can also include **comment** directives anywhere within the encryption envelope, including in the source code between the **begin** and **end** directives.

The rules that an encryption tool must follow to form a decryption envelope are somewhat more prescriptive. Groups of directives must appear in a specified order, even if the corresponding directives in the encryption envelope appeared in a different order or distributed among other directives, though not all groups are required in every decryption envelope. The layout of a decryption envelope is:

```
`protect begin_protected
author directives
license directives
encrypt agent directives
viewport directives
key block directives
data block directives
digest block directives
`protect end_protected
```

The author, license, and viewport directives are those that appear in the encryption envelope, if any. The **encrypt_agent** directive and optionally and **encrypt_agent_info** directive are included by the encryption tool. If a digital envelope is used, there is a group of key block directives for each encryption of the session key. The directives occur in the following order, with only the **key_keyname** directive being optional:

```
key_keyowner directive
key_keyname directive
key_keymethod directive
encoding directive
key_block directive
encoded cipher text for session key
```

The data block directives occur in the following order, with the **data_keyowner** and (optional) **data_keyname** directives only appearing if a digital envelope is not being used:

```
data_keyowner directive
data_keyname directive
data_method directive
encoding directive
```

```
data_block directive
encoded cipher text for source code
```

If a digital signature is used, the digest block directives occur in the following order, with only the **digest_keyname** directive being optional:

```
digest_keyowner directive
digest_keyname directive
digest_key_method directive
digest_method directive
encoding directive
digest_block directive
encoded cipher text for digest
```

2.5.1   Key Exchange

In our description of IP exchange so far we have assumed that the IP provider's encryption tool and the customer's decryption tool each have the required keys. What we have glossed over is how the tools get the keys. This is a very important topic, since protection of IP from disclosure relies on the security of the encryption and decryption keys. Should a key become known to an unauthorized party, the encrypted IP can be decrypted and disseminated. Normally, when encryption is used to secure communication between two parties, the parties are assumed to have an interest in the security of the encrypted messages and can be trusted to keep the keys secret. However, as we mentioned earlier, when an IP provider delivers a model to a customer, it is the customer's tool that is really the communicating party. The IP provider may not trust the customer not to look at the code or use it in some unauthorized way. A further complication is that the customer may have to provide his or her tool's key to an IP provider, creating an opportunity for the customer to copy the key and subsequently decrypt the code. Given these considerations, we can see that exchange of keys can be quite complicated. VHDL-2008 does not specify how keys should be exchanged; that is left to negotiation between IP providers, tool vendors, and customers. The following discussion, drawn from the VHDL standard, explores some of the issues.

Many applications that require secure exchange of keys rely on *public key infrastructure* (PKI). Parties to communication generate, or are given, key pairs for use with asymmetric ciphers. Each party keeps their private key secret, and publishes their public key, for example, in a directory. In order to establish that a public key does, in fact, belong to a given party, the public key is digitally signed by a trusted authority. The signed public key is represented in the form of a digital certificate, containing the key and the signature. The trusted authority is called a certification authority (CA). Many PKI systems have a hierarchy of CAs, allowing a certificate signed by a subordinate CA to be signed by a superior CA, allowing trust to be distributed hierarchically. One or more root CAs are required to be globally trusted.

Key exchange for IP protection may be built upon public key infrastructure. For example, a vendor of a decryption tool may embed a private key of a key pair in the tool and register the public key with a CA. The tool can then generate a key pair for the tool's user, keeping the private key secret and signing the public key with both the vendor's private key and the user's private key. This allows verification that the public key originates with the instance of the vendor's tool owned by the tool user. That public key may then be used by IP authors to provide IP for that use of that tool only. Similar mechanisms might also be employed within tools to allow exchange of private keys among tools without disclosure to the tools' user.

In addition to providing for secure key exchange, a decryption tool must take measures to ensure that stored keys are not disclosed to the tool user. If a tool user could read a tool's stored keys, the user could decrypt IP independently of the tool. One way of ensuring security of a tool's keys is for the tool to encrypt its key store using a secret key embedded in the tool in a disguised manner, and to provide for update and re-encryption of the secret key in case it is compromised.

2.6   VHDL Procedural Interface (VHPI)

VHPI is an application-programming interface (API) to VHDL tools. Using VHPI, a program can access information about a VHDL model during analysis, elaboration, and execution of the model. VHPI allows development of add-in tools, such as linters, profilers, code coverage analyzers, timing and power analyzers, and external models, among others. Use of the VHPI to develop such tools is quite complex, and is beyond the scope of this book. Instead, we will describe the way in which we can invoke VHPI programs as part of a VHDL simulation.

VHPI programs are divided into two classes: *foreign models* and *foreign applications*. A foreign model corresponds to an architecture or a subprogram decorated with the 'FOREIGN attribute. The VHPI program implements the behavior of the architecture or subprogram, respectively. A foreign application does not have a counterpart in the VHDL code. It is executed as part of simulation and performs application-specific processing. Both forms of VHPI program can use API calls to obtain information about the VHDL model, to react to changes in the simulation state, and to cause changes in the simulation state.

2.6.1   Direct Binding

If we are to instantiate a foreign model as part of a VHPI design, we need to identify where the VHPI program code is to be found. Typically, the provider of the foreign model would provide documentation listing the names of libraries and functions to which we should refer. The most straightforward method of referring to the VHPI program code is to provide the information in the value of the 'FOREIGN attribute in a form known as *direct binding*. For a foreign architecture, we write the attribute value in the following form:

```
"VHPIDIRECT object_lib_path elab_function exec_function"
```

The keyword **VHPIDIRECT** specifies standard direct binding, and must be written in uppercase. The *object_lib_path* is a host-dependent path and file name identifying the binary object library in the host file system. It can contain any characters; however, if a space character is required, we must precede it with a backslash character, and if a backslash character is required, we must double the backslash. The *elab_function* is the name of a function within the object library that performs elaboration for the foreign architecture. It is called to elaborate each instance of the foreign architecture during elaboration of the enclosing design. The *exec_function* is similarly the name of a function in the object library that performs simulation for the foreign architecture. It is called once for each instance of the foreign architecture during the initialization phase of simulation.

In the attribute value, we can substitute the keyword **null** for the object library path. In that case, the host system locates the object library in an implementation-dependent way. It might, for example, use an environment variable containing a list of pathnames. We can also substitute the keyword **null** for the elaboration function name if the foreign model does not require any action during elaboration. In both cases, the keyword **null** must be written in lowercase.

EXAMPLE 2.14   *Foreign processor core model*

Suppose a foreign model for a CPU32 processor core is provided in an object library called **cpu32.a** that we have installed in the directory **/usr/local/cpu32**. The elaboration and execution functions for a bus-functional version are named **cpu32_bf_elab_f** and **cpu32_bf_exec_f**, respectively. An entity and architecture that use standard direct binding for the bus-functional version are:

```
entity cpu32 is
  generic ( ... );
  port ( ... );
end entity cpu_32;

architecture bus_functional of cpu32 is
  attribute FOREIGN of bus_functional : architecture is
    "VHPIDIRECT /usr/local/cpu32/cpu32.a " &
    "cpu32_bf_elab_f cpu32_bf_exec_f";
begin
end architecture bus_functional;
```

The attribute value for standard direct binding for a foreign subprogram takes a similar form:

"VHPIDIRECT *object_library_path exec_function*"

In this case, the execution function name identifies a function that performs the action of the foreign subprogram. It is called whenever the foreign subprogram is called during simulation. For foreign subprograms, we can substitute the keyword **null** for the execution function name. In that case, the execution function name is taken to be the same as that of the foreign subprogram declared in the VHDL model, using the case of letters in the VHDL declaration.

EXAMPLE 2.15   *Foreign display subprograms*

Suppose we are given subprograms that show 7-segment display digits graphically on the screen during simulation. The subprograms are in the library **displaylib.a**, and include a function named **create_digit** and a procedure named **update_digit**. We can declare corresponding foreign subprograms in a package as follows:

```
package display_pkg is
  impure function create_digit (title : in string)
                                   return natural;
  attribute FOREIGN of create_digit : function is
    "VHPIDIRECT displaylib.a null";
  procedure update_digit (id : in natural;
                          val : in bit_vector(0 to 7));
  attribute FOREIGN of update_digit : procedure is
    "VHPIDIRECT displaylib.a null";
end package display_pkg;
```

2.6.2   Tabular Registration and Indirect Binding

An alternative way of identifying the VHPI program code for a foreign model is to use a *tabular registry*, which is a text file containing the identifying information. A tool can be supplied with any number of tabular registry files, each describing one or more foreign models or applications. The way in which we specify use of a tabular registry file is tool-dependent. It might, for example, involve use of a command-line option or an entry in an options-setting file. Each line of a tabular registry is an entry describing one foreign model, foreign application, or library of VHPI programs. The file can also contain comment lines, starting with characters "--", and blank lines.

A foreign architecture is described by a line of the following form in a tabular registry:

object_lib_name model_name **vhpiArchF** *elab_function exec_function*

The *object_lib_name* is a logical name for the binary object library containing the VHPI program code. The host system maps the logical name to a physical object library in some host-dependent way. The *model_name* is an identifier for the foreign architecture in the object library. Both the library logical name and the model name can be written as a normal identifier or, if non-standard characters are required, as an extended identifier delimited by backslash characters. The keyword **vhpiArchF** indicates that the line in the tabular registry describes a foreign architecture. It must be written using the combination of uppercase and lowercase letters shown here. The *elab_function* and *exec_function* are the names of the elaboration function and execution function, respectively, in the object library. They serve the same purpose as described in Section 2.6.1, and, in a similar way, the elaboration function name can be replaced by the keyword **null**.

Having described a foreign architecture in a tabular registry file, we can specify a 'FOREIGN attribute in the form of an *indirect binding* to use the foreign architecture for a VHDL architecture. This form of attribute value is:

"**VHPI** *object_lib_name model_name*"

The *object_lib_name* and *model_name* identifiers must correspond to the library logical name and model name identifiers specified in an entry in a tabular registry. The foreign architecture described in that entry is used for each instance of the VHDL architecture decorated with the attribute.

EXAMPLE 2.16   *Foreign processor core model using indirect binding*

Suppose the provider of the CPU32 processor core model described in Example 2.14 also provides a tabular registry file for binding the bus-functional model. The file contains the following entry:

cpu32lib \cpu32-bf\ **vhpiArchF** cpu32_bf_elab_f cpu_bf_exec_f

We decorate the architecture with the 'FOREIGN attribute using indirect binding for the bus-functional model:

```
architecture bus_functional of cpu32 is
  attribute FOREIGN of bus_functional : architecture is
    "VHPI cpu32lib \cpu32-bf\";
begin
end architecture bus_functional;
```

Tabular registration and indirect binding for a foreign subprogram are similar. An entry in a tabular registry file for a foreign procedure takes the form:

object_lib_name model_name **vhpiProcF** **null** *exec_function*

and for a foreign function:

object_lib_name model_name **vhpiFuncF** **null** *exec_function*

In both cases, the *object_lib_name* and *model_name* serve the same purpose as for a foreign architecture, and the *exec_function* is the name of the function in the object library that implements the subprogram's actions. The function name can be replaced by the keyword **null**, in which case the execution function is taken to be the same as the model name. The 'FOREIGN attribute value for indirect binding to a foreign subprogram is the same as that for indirect binding to a foreign architecture, namely,

"**VHPI** *object_lib_name model_name*"

The library name and model name are used in the same way to locate the tabular registry entry for the foreign subprogram.

EXAMPLE 2.17 *Foreign display subprograms using indirect binding*

The provider of the display subprograms described in Example 2.15 might provide a tabular registry file for the subprograms including the following entries:

```
displaylib create_digit vhpiFuncF null null
displaylib update_digit vhpiProcF null null
```

The second **null** in each entry indicates that the execution function names for the subprograms are the same as the foreign model names, namely, **create_digit** and **update_digit**. We declare the foreign subprograms and use indirect binding in the 'FOREIGN attribute values as follows:

```
package display_pkg is
  impure function create_digit (title : in string)
                                return natural;
  attribute FOREIGN of create_digit : function is
    "VHPI displaylib create_digit";
  procedure update_digit (id : in natural;
                          val : in bit_vector(0 to 7));
  attribute FOREIGN of update_digit : procedure is
    "VHPI displaylib update_digit";
end package display_pkg;
```

2.6.3 Registration of Applications and Libraries

We can use the tabular registration feature described in Section 2.6.2 to describe a VHPI application to be run as part of a simulation. A line in the file for a foreign application takes the form:

```
object_lib_name application_name vhpiAppF reg_function null
```

The *object_lib_name* is a logical name identifying the binary object library containing the program code, and the *application_name* is an identifier for the foreign application in the object library. The rules for these names are the same as those for names identifying foreign models. Thus, they can be written as normal identifiers or extended identifiers. The keyword **vhpiAppF** indicates that the line in the tabular registry describes a foreign application and must be written using the combination of uppercase and lowercase letters shown here. The *reg_function* is the names of a function in the object library that is called at the start of simulation, before elaboration or initialization, to initialize the state of the foreign application. This is all the information we need to supply to the tool to include a foreign application in a simulation. The registration function performs any further application-specific operations required.

EXAMPLE 2.18   *Registration of a power-estimation application*

A third-party tool supplier might provide a tool for estimating dynamic power consumption based on activity during simulation of a model. The tool's program code is installed in a binary object library in the host file system, with a logical name **power-estlib** mapping to the library file. The application is named **powerest**, and the registration function in the library is called **powerest_reg_f**. The supplier provides a tabular registry file with the following contents:

```
-- VHPI tabular registry for the PowerEst foreign application.
-- Map library name powerestlib to the pathname for the
-- powerestlib.a file in your installation.

powerestlib powerestlib vhpiAppF powerest_reg_f null
```

We invoke the simulator with a command-line option identifying this tabular registry file to include the power estimator tool in a simulation.

The final form of entry in a tabular registry file describes a library of VHPI programs, including foreign models or applications. The form of the entry is:

```
object_lib_name null vhpiLibF reg_function null
```

As before, the *object_lib_name* is a logical name identifying the binary object library containing the program code. The *reg_function* is the names of a function in the object library that is called at the start of simulation. It uses the VHPI API to register each foreign model or application. This form of registration is convenient when a large suite of VHPI programs is provided.

Chapter 3

Type System Changes

VHDL is a strongly-typed language, which means that every object has a specified type, specification of types is explicitly stated, and correct use of types is required. One rationale for strong typing is that it helps tools detect errors early in the design process, usually during analysis, rather than later during elaboration or execution. This helps designers avoid the escape of bugs into products. Another rationale is that it provides extra information to an analyzer, so it can generate code optimized for a particular use of data. There is a trade-off in supporting these benefits. The type rules can seem somewhat restrictive or burdensome to the designer. In particular, rules that make it easier for a tool to implement language constructs can make it harder for a designer to write code expressing their intent.

In this chapter, we describe two ways in which VHDL-2008 changes the type system to relax some of the type rules. Both changes deal with rules relating to elements of composite types. The changes imply that tools must do more work to check correctness of designs and generate corresponding code for simulation. However, they remove restrictions that designers have found burdensome in earlier versions of VHDL.

3.1 Unconstrained Element Types

VHDL provides two kinds of composite types, namely, arrays and records. Each contains elements: all of the same type, in the case of array elements; and of heterogeneous types, in the case of record elements. In earlier versions of VHDL, the element types for arrays and records all had to be constrained, meaning the size of any elements that were arrays had to be fixed. So, for example, if we had a type that was an array of arrays, we could leave the outer array size unspecified, but the element array size had to be fixed. This restriction has long been an impediment, but it was not a simple matter to change. Hence, successive revisions of the VHDL standard left the restriction in place. In VHDL-2008, however, considerable effort has been invested in revising the type rules to lift the restriction. We describe the new rules in this section.

3.1.1 Composite Types

In order to fully understand the rules, we first review some terminology used in the VHDL standard. A *type* in VHDL just specifies a set of values. A *subtype* is a subset of values from a type, determined by a *constraint*. Those values that meet the constraint are in the subtype, and those that don't meet the constraint are not in the subtype. The type from which values of a subtype are drawn is called the *base type* of the subtype. Note that a subtype need not be a proper subset of a type. The constraint may be vacuous, allowing any value from the type to be in the subtype. Thus, a type is considered to be a subtype of itself.

For an array type, each value is an indexed collection of elements, each of the same subtype, called the *element subtype*. An array value has one or more indices, the number of which determine the dimensionality of the array. Each index comes from an *index subtype*, which must contain only discrete scalar values, such as integers or values of an enumeration type. Thus, a one-dimensional array type, commonly called a vector, has a single index subtype; a two-dimensional array type, commonly called a matrix, has two distinct index subtypes; and so on.

A particular value or object of an array type has an *index range* for each index position. The index range has a left bound, a right bound, and a direction; it also determines the number of elements in the array value. Note that an index range is a distinct concept from an index subtype; it is a property of an array value or an array object, whereas an index subtype is a property of an array type. Keeping these concepts distinct in our minds will help us make sense of the type rules. The connection between the two is that, if an array value or object is of an array type, the bounds of each index range must belong to the corresponding index subtype of the array type.

When we declare an array type, we are effectively declaring an array subtype. There are two forms of array type declaration. One form, called a *constrained array* declaration, specifies index ranges, for example:

```
type A1 is array (natural range 0 to 7) of bit;
```

In this example, we are declaring an anonymous array base type that has **natural** as its index subtype and **bit** as its element subtype. The name **A1** denotes a subtype of that anonymous base type, with the *index constraint* that the index range for values of the subtype must have a left bound of 0, a right bound of 7, and an ascending direction.

The other form of array type declaration was called an unconstrained array declaration in earlier versions of VHDL, but in VHDL-2008, it is called an *unbounded array* declaration. An example is:

```
type A2 is array (natural range <>) of bit;
```

Here, we are declaring **A2** to be a base type that has **natural** as its index subtype and **bit** as its element subtype. We can treat **A2** as a subtype with vacuous constraint; that is, values of the subtype can have any index range, of either direction, provided the bounds are values of subtype **natural**.

VHDL-2008 defines the terms "unconstrained" and "constrained" somewhat differently from earlier versions of VHDL, since the situation is somewhat more involved. In VHDL-2008, an array subtype is *unconstrained* if it has no index constraints, and the element subtype is either not a composite type or is an unconstrained type. An unconstrained type has no constraints anywhere in its structure where a constraint could apply. An array subtype is *fully constrained* if it has index constraints for all of its indices, and the element subtype is either not a composite type or is a fully constrained type. A fully constrained type has constraints everywhere in its structure where a constraint could apply. Together, these two categories do not cover all array subtypes, as there are those with index constraints but not fully constrained element subtypes, and those with no index constraints but full constrained element subtypes. These "in between" subtypes are called *partially constrained*, and have constraints in some places but not in others where a constraint could apply.

Here are some examples to illustrate the categories of array types. We will also use these types in further examples in Section 3.1.2. First, array types with non-composite elements are either unconstrained or fully constrained. Thus

```
type M_unconstrained is
  array (natural range <>, natural range <>) of bit;
```

is unconstrained, and

```
type M_fully_constrained is
  array (natural range 0 to 7, integer range -1 to 1) of bit;
```

is fully constrained. Note that every index of any array (but not necessarily of its elements) must be constrained or not. Thus, we could not legally write

```
type M_illegal is
  array (natural range <>, integer range -1 to 1) of bit;
```

Now, if we use **M_unconstrained** as the element subtype in an unbounded array definition:

```
type A_unconstrained is
  array (character range <>) of M_unconstrained;
```

the subtype defined is unconstrained. This was illegal in earlier versions of VHDL, but is now legal in VHDL-2008. If we use **M_fully_constrained** in a constrained array definition:

```
type A_fully_constrained is
  array (character range 'A' to 'Z') of M_fully_constrained;
```

the subtype defined is fully constrained. This was legal in earlier versions of VHDL, and remains legal in VHDL-2008. We can define partially constrained array subtypes as follows:

```
type A1_partially_constrained is
  array (character range 'A' to 'Z') of M_unconstrained;
type A2_partially_constrained is
  array (character range <>) of M_fully_constrained;
```

An object of subtype **A1_partially_constrained** must have 'A' to 'Z' as its index range, but the index ranges of each element are not specified. This was illegal in earlier versions of VHDL. An object of subtype **A2_partially_constrained** does not have its index range specified, but the index ranges of each element must be 0 to 7 and −1 to 1, respectively. This was previously legal, and remains legal in VHDL-2008.

VHDL-2008 also makes similar extensions to the type rules for records. For a record type, a value is a collection of elements, each identified by an element name and each of a specified element subtype. The element subtypes for different elements need not be the same. Moreover, in VHDL-2008, the element subtypes need not be constrained, as they were in earlier versions of VHDL. Just as we did for array subtypes, we say that a record subtype is unconstrained if each element subtype is either not a composite type or is an unconstrained type. A record subtype is fully constrained if each element subtype is either not a composite type or is a fully constrained type. A record subtype that is neither unconstrained nor fully constrained is partially constrained.

Again, here are some examples illustrating the categories for record types. First, a record type with non-composite elements is fully constrained:

```
type R_non_composite_elements is record
  e1 : bit;
  e2 : integer;
end record R_non_composite_elements;
```

It cannot be unconstrained, since there is no place in the record structure for a constraint to apply. An example of a record type declaration defining an unconstrained record subtype is:

```
type R_unconstrained is record
  e1 : A_unconstrained;
  e2 : M_unconstrained;
  e3 : bit;
end record R_unconstrained;
```

In this case, the first two element subtypes are unconstrained and the third is non-composite, so the subtype is unconstrained. A record type declaration with fully constrained elements is:

```
type R_fully_constrained is record
  e1 : A_fully_constrained;
  e2 : M_fully_constrained;
  e3 : bit;
end record R_full_constrained;
```

This defines a fully constrained record subtype. The declaration:

```
type R1_partially_constrained is record
  e1 : A_unconstrained;
  e2 : M_fully_constrained;
  e3 : bit;
end record R1_partially_constrained;
```

defines a partially constrained record subtype, since the composite element subtypes are neither all unconstrained nor all fully constrained. Similarly, the declarations:

```
type R2_partially_constrained is record
   e1 : A1_partially_constrained;
   e2 : M_fully_constrained;
   e3 : bit;
end record R1_partially_constrained;

type R3_partially_constrained is record
   e1 : A2_partially_constrained;
   e2 : M_unconstrained;
   e3 : bit;
end record R3_partially_constrained;
```

both define partially constrained record subtypes, for the same reason.

3.1.2   Subtype Indications and Constraints

Now that we have seen how to declare unconstrained, partially constrained, and fully constrained subtypes using type declarations, we can turn to the way in which we specify constraints in subtype declarations and other places. In earlier versions of VHDL, the only kind of constraint we could apply to a composite subtype was an index constraint to specify index ranges for an otherwise unconstrained array type. All of the array elements and subelements of the array type had to be constrained. The situation is different in VHDL-2008, since we can have array and record types in which some element or subelement subtype is an array subtype without an index constraint. We may want to specify index ranges at various levels in a composite subtype's nested structure. We will show how the rules for specifying constraints are extended in VHDL-2008 for that purpose.

There are numerous places in VHDL where we can specify a subtype using a *subtype indication*, including in subtype declarations; declarations of elements of composite and other types; declarations of constants, signals, variables and files; declarations of generics, ports and parameters; and so on. In each case, the subtype indication takes the form of the name of a type or subtype followed by a constraint that limits the values that are in the subtype. We will just use subtype declarations to illustrate the various forms of subtype indication, but the same rules apply in other places. The types that we refer to in the following examples are all defined in Section 3.1.1.

If the type we are constraining is an array subtype with unspecified index ranges, we can include an index constraint, as we did in earlier versions of VHDL, for example:

```
subtype S1 is A_unconstrained('x' to 'z');
```

In this case, since the element subtype for **A_unconstrained** was also unconstrained, it remains unconstrained in S1. Thus, S1 is a partially constrained subtype. If we write

```
subtype S2 is A2_partially_constrained('c' downto 'a');
```

the subtype S2 is fully constrained, since the element subtype of **A2_partially_ constrained** is fully constrained.

Now suppose we want to define a subtype of **A_unconstrained** specifying index ranges for the top-level array and also for the elements. We can do this as follows:

```
subtype S3 is A_unconstrained('x' to 'z')(0 to 7, 31 downto 16);
```

The subtype S3 is fully constrained. Values or objects of this subtype must have three elements indexed from 'x' to 'z', and each element must be a matrix indexed from 0 to 7 in one dimension and from 31 down to 16 in the other. If we want to define a partially constrained subtype, specifying index ranges for the elements but leaving the index range at the top level unspecified, we can use the reserved word **open** in place of an index range, for example:

```
subtype S4 is A_unconstrained(open)(0 to 7, 31 downto 16);
```

Values or objects of this subtype can have any index bounds and direction, provided each element is a matrix indexed from 0 to 7 in one dimension and from 31 down to 16 in the other. We can also use this notation to specify index ranges for elements of a sub-type that already has a constraint on the top-level indices, for example:

```
subtype S5 is
  A1_partially_constrained(open)(0 to 7, 31 downto 16);
```

In this case, the subtype **A1_partially_constrained** specifies an index range of 'A' to 'Z', but leaves the index ranges for the elements unconstrained. In the subtype declaration for S5, the use of **open** for the top-level constraint indicates that we leave the existing constraint, and skip over to the elements, for which we do specify index ranges.

As in earlier versions of VHDL, we can't specify index ranges in a subtype indication if the subtype already has an index constraint at the specified position. So the following would be illegal:

```
subtype S6 is A1_partially_constrained('è' to 'ë'); -- illegal
```

since the constraint 'è' to 'ë' conflicts with the constraint 'A' to 'Z' specified in the subtype A1_partially_constrained.

If the type we are constraining is a record type or subtype, we can specify index constraints for the element subtypes. We need to indicate which record element is being constrained. An example showing the notation we use is:

```
subtype S7 is R_unconstrained(e1('A' to 'F'));
```

The subtype S7 is partially constrained with an index range of 'A' to 'F' specified for the element e1, but no index ranges specified for the elements of e1 or for the element e2. We could further constrain S7 as follows:

```
subtype S8 is S7(e1(open)(15 downto 0, 7 downto 0));
```

In this case, since e1 already includes an index constraint, we skip over it and specify index ranges for the elements of e1. Again, we leave the element subtype for e2 unconstrained. We can specify constraints for multiple elements as follows:

```
subtype S9 is S7(e1(open)(15 downto 0, 7 downto 0),
                e2(1 to 3, 1 to 3) );
```

Now, having specified constraints in all places where a constraint can apply, the subtype S9 is fully constrained.

This notation for specifying constraints for subelements of composite types has sufficient generality to work for any arbitrary nesting. We just need to ensure that the way we list constraints follows the hierarchy of nesting. At any given level, if the element subtype is an array subtype, we either specify one or more index ranges in parentheses or use the reserved word **open** to skip the level without specifying any index ranges. If the element subtype at the given level is a record subtype, we write one or more element names, and for each, we write constraints for the named element subtype. For example, given the following type declarations:

```
type T1 is array (integer range <>) of T;
type T2 is array (integer range <>, integer range <>) of T;
type T3 is record
  e1 : T1;
  e2 : T2;
end record T3;
type T4 is array (integer range <>) of T3;
type T5 is array (integer range <>, integer range <>) of T4;
```

We can write a fully constrained subtype declaration as follows:

```
subtype S10 is
  T5(1 to 4, 0 to 9)
      (3 downto 0)
        (e1(9 to 99),
          e2(-1 to 1, -1 to 1) );
```

3.1.3   Use of Composite Subtypes

In earlier versions of VHDL, there were rules specifying where we had to use a constrained subtype and where we could use an unconstrained subtype. These rules have been modified in VHDL-2008 to reflect the new category of partially constrained subtypes. In general, where previously we had to use a constrained subtype, we must now use a fully constrained subtype. Where previously we could use an unconstrained or a constrained subtype, we can now use an unconstrained, partially constrained, or fully constrained subtype. In earlier versions of VHDL, the rules for determining the index ranges for array objects were somewhat unclear. VHDL-2008 clarifies the rules, and extends them to deal with determining the index ranges for arrays that are elements or subelements of larger composite objects. We will go through the cases covered by these rules and use examples to show how they apply. In each case, it is important to keep in mind the distinction between an index subtype for an array type and an index range of a value or object. The cases we are discussing here deal with the way the index ranges for values or objects are determined from various subtypes in different ways.

Variable and Signal Declarations

The first case deals with variable and signal declarations, and includes arrays that are whole objects and arrays that are elements or subelements of larger objects. In this case, the subtype of the variable or signal must be a fully constrained subtype, since a tool must be able to determine the size of the object. For example, if we have a type declared as:

```
type signed_matrix is array (1 to 3, 1 to 4) of signed;
```

we must constrain the element type in order to declare a variable:

```
variable v : signed_matrix(open)(7 downto 0);
```

This gives a fully constrained subtype for the variable. A tool can then determine that the variable needs $3 \times 4 \times 8 = 96$ scalar subelements. The index ranges for the array variable are taken from the fully constrained subtype, and are 1 to 3 for the first dimension and 1 to 4 for the second dimension. The index range for each element array are also taken from the fully constrained subtype. So each element has the index range 7 down to 0.

Constant Declarations

The second case deals with constant declarations, in which we can use unconstrained, partially constrained, and fully constrained subtypes. We must provide a value for a declared constant, and the value must belong to the subtype of the constant. The actual index ranges for the constant are determined jointly from the subtype and from the index ranges of the value. If the subtype includes a constraint that specifies index ranges at any given position, those index ranges are used. If the subtype leaves the index ranges undefined at any given position, the index range for that position is the corresponding index range from the value. In either case, whether the index range be specified by a constraint or determined from the value, each element of the constant is the element in the same left-to-right position in the initial value.

To illustrate the rule for constants, suppose we have the following declarations:

```
type A is array (1 to 3) of bit_vector;
constant C : A := ("0100", "1101", "0010");
```

The subtype A specifies an index range of 1 to 3 for the top-level array, so that is the index range used. However, the element subtype of A is unconstrained, so the index range for each element comes from the initial value's elements. Those elements are all bit strings of type bit_vector, with index range 0 to 3. The subelements of the constant value are thus:

```
C(1)(0) = '0'    C(1)(1) = '1'    C(1)(2) = '0'    C(1)(3) = '0'
C(2)(0) = '1'    C(2)(1) = '1'    C(2)(2) = '0'    C(2)(3) = '1'
C(3)(0) = '0'    C(3)(1) = '0'    C(3)(2) = '1'    C(3)(3) = '0'
```

Now suppose we declare a further constant as:

```
constant C1 : A(open)(7 downto 4) := C;
```

In this case, the subtype for the constant C1 specifies index ranges at both levels: 1 to 3 for the top level and 7 down to 4 for the element level. Since the elements and sub-elements are assigned from the initial value in left-to-right order, the values are:

```
C1(1)(7) = '0'   C1(1)(6) = '1'   C1(1)(5) = '0'   C1(1)(4) = '0'
C1(2)(7) = '1'   C1(2)(6) = '1'   C1(2)(5) = '0'   C1(2)(4) = '1'
C1(3)(7) = '0'   C1(3)(6) = '0'   C1(3)(5) = '1'   C1(3)(4) = '0'
```

One point to note is that, since the intial value is converted to the subtype of the constant, the initial value doesn't need to have exactly the same index range as the constant, providing the length matches. For example, given the following declarations

```
type A2 is array (2 downto 0) of bit_vector;
constant C2 : A2 := C;
```

the index range of C2 is 2 down to 0, but the index range of the initial value C is 1 to 3. Elements of C are used to initialize C2 from left to right.

Attribute Specifications

The third case in the rules for determining index ranges deals with attribute values in attribute specifications, and is similar to the case of constant declarations. Thus, if the subtype in an attribute specification is an array subtype or includes array elements, the attribute value must belong to the subtype of the constant, and the actual index ranges are determined jointly from the subtype and from the index ranges of the specified value. In a sense, an attribute specification defines a constant value that decorates the named item. As an example, if we declare an attribute of a composite type:

```
type string_vector is array (positive range <>) of string;
attribute key_vector : string_vector;
```

we can decorate an item with the attribute as follows:

```
attribute key_vector of e : entity is
  ("66A6D 7DF3A 88CE1 8DEEB", "012BD 2BEE9 98634 93FE1");
```

Since the subtype for the attribute specifies index ranges in neither the top-level nor the element position, the corresponding index subtypes are used to determine the index ranges for the attribute value, giving the ranges 1 to 2 for the top level and 1 to 23 for each element.

Allocated Objects

The fourth case deals with allocation of objects using **new**. In this case, the allocator determines the index ranges of the allocated object. If we write an allocator with just a subtype indication, it must specify a fully constrained subtype, and the index ranges are taken from that subtype. For example, given the following declarations:

```
type RV is record
  v1 : bit_vector;
  v2 : time_vector;
end record RV;
type RV_ptr is access RV;
variable p : RV_ptr;
```

we can write an allocator using a subtype indication:

```
p := new RV_record(v1(0 to 23), v2(0 to 23));
```

The subtype indication specifies index ranges of 0 to 23 for both elements, so they are used for the allocated object. The object is then initialized with the default initial value.

On the other hand, if we write an allocator with a qualified expression, the value in the expression is converted to the named subtype (see Sections 9.3 and 9.4), and that determines the index ranges for the allocated object. Where that subtype specifies index ranges, they are used; and where no index range is specified, an index range is determined from the corresponding index subtype. For example, given the preceding declarations, we can write an allocator:

```
p := new RV_record'(v1 => "010", v2 => (2 ns, 4 ns, 6 ns));
```

Since the subtype **RV_record** does not specify any index ranges, the index subtypes for the record elements are used to determine index ranges for the allocated value. For each element, the index subtype is **natural**, so the index ranges are 0 to 2.

Interface Objects

The fifth and final case deals with interface objects, namely, generic constants, ports, and parameters. The declaration of a formal interface object includes a subtype indication, which may define an unconstrained, partially constrained, or fully constrained subtype. For each index position in the formal, whether it be at the top level of an array or a sub-element within the composite structure of the formal, there may or may not be a corresponding index range defined by an index constraint in the subtype. We call the index range specified in the subtype, if defined, the *subtype index range* corresponding to a given index position in the formal. We will now see how to determine the index range for each index position of the formal. There are several subcases, depending on the declaration of the formal, the actual value or object associated with the formal, and the way in which the association is written in the relevant generic map, port map, or parameter list.

In the first subcase, the subtype index range is defined by an index constraint in the subtype. The index range for the formal is taken from that index constraint. As an example, suppose we declare a port of an entity as follows:

```
entity ent1 is
  port ( p : out std_logic_vector(0 to 31) );
end entity ent1;
```

Regardless of the actual signal associated with the port in an instance of the entity, the formal port takes its index range from the subtype, since the subtype defines the index range. Thus, in an architecture, we can reference the index range as follows:

```
architecture a of ent1 is
begin
  process is
  begin
    . . .
    for i in 0 to 31 loop
      p(i) <= ...
    end for;
    . . .
  end process;
end architecture a;
```

We can declare a signal and associate it with the port of an instance of ent1:

```
signal s : std_logic_vector(63 downto 32);
. . .

inst1 : entity work.e(a)
  port map ( p => s );
```

The fact that the actual signal has a different index range does not affect the index range for the formal port. All it means is that s(63) is associated with p(0), s(62) with p(1), and so on.

This subcase also applies to subelements of interface objects. For example, suppose we have a type declared as:

```
type byte_vector is
  array (natural range <>) of bit_vector(7 downto 0);
```

We might define a function as follows:

```
function reduce ( v : byte_vector ) return bit is
  . . .
begin
  for i in v'range loop
    for j in 7 downto 0 loop
      . . .
    end loop;
  end loop;
  return ...;
end function reduce;
```

In any call to **reduce**, the index range for each element of **v** is 7 down to 0, taken from the subtype **byte_vector**, regardless of the index range at the top level of **v**.

In the second subcase, the subtype index range is undefined, but the array is associated using subelement association. That means the association list uses named association to divide the formal into separate elements or slices and associates each element or slice with a separate actual value or object. The index range for the formal is then determined from the index values used in those formal element names or slices. The smallest index value used is the low bound of the index range of the formal, and the largest index value used is the high bound. The direction of the index range is the direction of the corresponding index subtype taken from the subtype of the formal. To illustrate, we can revise our earlier example of an entity's port:

```
entity ent2 is
  port ( p : out std_logic_vector );
end entity ent2;
```

In this example, we don't know the index range for **p** within the architecture body, so we would have to refer to it using attributes, such as **p'range**. If we write an instance of the entity as follows:

```
inst2 : entity work.ent2(a)
  port map ( p(11) => s1, p(12 to 15) => sv );
```

the index values in the formal element and slice names are used to determine the index range for the formal **p** for this instance. The smallest value is 11, and the largest value is 15. The direction for the index range is ascending, since the index subtype for **std_logic_vector** is **natural**, which is ascending. Thus, the index range for **p** is 11 to 15. Note that the direction for the index range for the port is determined by the index subtype of the port, not by the direction of the range in a slice name in the port map. Had we written the above instantiation as:

```
inst2 : entity work.ent2(a)
  port map ( p(15) => s1, p(14 downto 11) => sv ); -- illegal
```

the slice name in the port map would be in error. VHDL requires that the direction of the range in a slice name match the direction of the index range of the array being sliced.

As before, this subcase also applies to subelements of interface objects. For example, given a type

```
type bv_pair is array (1 to 2) of bit_vector;
```

and a port declared in an entity as:

```
entity ent3 is
  port ( p : in bv_pair );
end entity ent3;
```

we can write an instance of the entity:

```
signal s1, s2 : bit;
signal sv1, sv2 : bit_vector(4 to 7);

inst3 : entity work.ent3
  port map ( p(1)(0) => s1, p(1)(1 to 4) => sv1,
             p(2)(0) => s2, p(2)(1 to 4) => sv2 );
```

Here, the index range for the top level of **p** is determined from the subtype, as in the first subcase. However, for the elements, the subtype index range is not defined, so the index range for **p** comes from the formal element and slice names. Combining these effects, the index ranges for **p** are 1 to 2 for the top level, and 0 to 4 for the elements. Note that the index range determined for the two elements **p(1)** and **p(2)** must be the same. It would be illegal to write the instance as:

```
signal s1, s2 : bit;
signal sv1, sv2 : bit_vector(4 to 7);

inst3 : entity work.ent3
  port map ( p(1)(0) => s1, p(1)(1 to 4) => sv1,
             p(2)(15) => s2, p(2)(11 to 14) => sv2 ); -- illegal
```

since that would imply two different index ranges: 0 to 4 for **p(1)** and 11 to 15 for **p(2)**. An array must have the same index ranges for all elements.

In the third subcase, the subtype index range is undefined, but the array is associated as a whole. There are no index values or slice values to identify the index bounds for the formal. Instead, the index range is determined from the corresponding index range of the actual or from any conversion functions or type conversions that appear in the association between the actual and the formal. We need to consider the various sub-subcases.

The first sub-subcase is a simple association involving no type conversions or conversion functions in the association between actual and formal. In this sub-subcase, the index range for the formal is taken from the corresponding index range of the actual. For example, given our entity declaration with an unconstrained port, as before:

```
entity ent2 is
  port ( p : out std_logic_vector );
end entity ent2;
```

we might write an instance as follows:

```
signal s12 : std_logic_vector(15 downto 4);
...

inst4 : entity work.ent2(a)
  port map ( p => s12 );
```

In this example, the index range of the formal is not defined, and the association with the actual provides no index values to use. So the formal takes its index range, 15 down to 4, from the associated actual signal s12.

Again, this rule applies to arrays that are subelements of interface objects. Returning to our entity ent3 with a port of type **bv_pair**, we can instantiate it as:

```
signal sv1, sv2 : bit_vector(0 to 7);

inst5 : entity work.ent3
  port map ( p(1) => sv1, p(2) => sv2 );
```

As before, the index range for the top level of **p** is determined from the subtype. However, for the elements, the subtype index range is not defined and the associations do not provide index values. Thus, the index range for the elements come from the actuals. Combining these effects, the index ranges for **p** are 1 to 2 for the top level, and 0 to 7 for the elements.

The second sub-subcase arises for an interface object of mode **in, inout,** or **linkage**, when the association with an actual includes a type conversion or conversion function applied to the actual. In this sub-subcase, the index range for the formal comes from the result of the conversion. This requires that the conversion define the corresponding index ranges. For a type conversion, the named type must be a subtype that includes a constraint defining the relevant index ranges, and for a conversion function, the result subtype must similarly define the relevant index ranges. To illustrate, suppose we have an entity with an unconstrained **in**-mode port, declared as follows:

```
type signed_vector is (natural range <>) of signed;
...

entity ent4 is
  port ( x : in signed_vector );
end entity ent4;
```

We can associate a signal of type **integer_vector** with the port, provided we apply a conversion function. However, the function must specify index ranges in its result subtype, since the port subtype has no index ranges specified. A legal example is:

```
subtype iv3 is integer_vector(1 to 3);
subtype sv3 is signed_vector(1 to 3)(31 downto 0);
signal iv : iv3;
function cvt3 ( v : iv3 ) return sv3;
...

inst6 : entity work.ent4
  port map ( x => cvt3(iv) );
```

Since the result subtype of the conversion function specifies the index ranges 1 to 3 at the top level and 31 down to 0 at the element level, those are the ranges used for the formal in the instance. Had we written a conversion function:

```
function cvt ( v : integer_vector ) return signed_vector;
```

we would not be able to use it in the same way, since it does not specify the index ranges to be used for the formal port in the instance. Similar arguments apply to type conversions. For example, with the following declarations:

```
type unsigned_vector is (natural range <>) of unsigned;
subtype uv3 is unsigned_vector(1 to 3)(31 downto 0);
signal uv : uv3;
```

we could write the following instance of entity ent4:

```
inst7 : entity work.ent4
  port map ( x => sv3(uv) );
```

since the subtype named in the type conversion specifies the index ranges 1 to 3 at the top level and 31 down to 0 at the element level, whereas the following would be illegal:

```
inst8 : entity work.ent4
  port map ( x => signed_vector(uv) );
```

The third sub-subcase is similar to the second. It arises for an interface object of mode **out**, **buffer**, **inout**, or **linkage**, when the association with an actual includes a type conversion or conversion function applied to the formal. For a type conversion, the named type must be a subtype that includes a constraint defining the relevant index ranges, and these are used for the formal. For a conversion function, the parameter subtype must similarly define the relevant index ranges, and these are used for the formal.

Note that both the second and the third sub-subcases deal with interface objects of modes **inout** and **linkage**. This mirrors the possibility of including type conversions or conversion functions in both the formal and actual parts of the association between formal and actual. If that occurs, both sub-subcases apply, and the index ranges determined must agree.

Summary: Determining Array Index Ranges

Since this case analysis of the way in which index ranges are determined is complex and multi-level, we'll summarize it here. For an array object, including a subelement array of a larger composite object, we determine each index range as follows:

1. For a declared signal or variable: The index range comes from the object's subtype, which must be fully constrained.

2. For a declared constant: If the constant's subtype defines the index range, that index range is used; otherwise, the index range comes from the corresponding index range of the constant's initial value.

3. For an attribute value: The index range comes from the attribute's subtype and the specified value, in same way as case 2.

4. For an allocated object: If the allocator is in the form of a subtype indication, the index range comes from the specified subtype, which must be fully constrained. Otherwise, the allocator is in the form of a qualified expression, and the index range

comes from the index range in the qualified expression's subtype, if defined, or from the index subtype in the qualified expression's subtype otherwise.

5. For a formal interface object, there are three subcases:

    a. If the formal's subtype defines the index ranges, they are used.

    b. If the formal's subtype does not define the index ranges, and subelement association is used to specify index values for the formal, then the index range uses the smallest and largest index values as the bounds, and gets the direction from the index subtype of the formal's subtype.

    c. If the formal's subtype does not define the index ranges, and the association with the actual does not specify index values for the formal, then there are three sub-subcases:

        *   If there are no conversions involved in the association, then the index range comes from the actual object.

        *   If the formal is of mode **in**, **inout**, or **linkage**, and there is a conversion in the actual part of the association, then the index range comes from the conversion's result subtype, which must define a corresponding index range.

        *   If the formal is of mode **out**, **buffer**, **inout**, or **linkage**, and there is a conversion in the formal part of the association, then the index range comes from the type conversion's subtype or conversion function's parameter subtype, as appropriate, and that subtype must define a corresponding index range.

        For a formal of mode **inout** or **linkage**, if conversions are used in both formal and actual parts, they must both define the same index ranges.

Type Conversions

Now that we've covered the rules dealing with the way in which index ranges are determined for composite objects, we can turn to some further uses of composite subtypes. One use is as the target of a type conversion. VHDL-2008 makes further changes to the rules for type conversions (see Section 9.4), but we will focus on the rules relating to index ranges here. When we convert the type of an array object to a target array subtype, we produce an array with the same element values, but with different index ranges. If the target subtype defines index ranges at a given index position, we use those index ranges. On the other hand, if the target subtype leaves the index ranges undefined, we determine index ranges for the result based on the index subtype at that position. The index range starts at the leftmost value of the index subtype and has the same direction, ascending or descending, as the index subtype. The right bound is then determined by the required size for the index range.

EXAMPLE 3.1 *Type conversions between signed and unsigned vectors*

Suppose we have two unconstrained types declared as:

```
type unsigned_vector is (natural range <>) of unsigned;
type  signed_vector is (natural range <>) of  signed;
```

and subtypes declared as:

```
subtype unsigned_vector3 is
        unsigned_vector(1 to 3);
subtype unsigned_byte_vector is
        unsigned_vector(open)(7 downto 0);
subtype unsigned_byte_vector3 is
        unsigned_vector(1 to 3)(7 downto 0);
```

Given a signal:

```
signal s : signed_vector(1 to 3)(7 downto 0);
```

the conversion:

```
unsigned_vector3(s)
```

yields an array indexed from 1 to 3 at the top level and from 0 to 7 at the element level. The top-level index range is specified in the target subtype. The element-level index range is determined from the index subtype **natural**, starting from 0 and ascending for eight elements. Alternatively, the conversion:

```
unsigned_byte_vector(s)
```

yields an array indexed from 0 to 2 at the top level and from 7 down to 0 at the element level. In this case, the target subtype does not specify an index range and the top level, so the top-level index range is determined from the top-level index subtype. The element-level index range comes from that specified in the target subtype. Finally, the conversion:

```
unsigned_byte_vector3(s)
```

yields an array indexed from 1 to 3 at the top level and from 7 down to 0 at the element level, since both index ranges are specified in the target subtype.

Alias Declarations and Subtype Attributes

Another place where we can use a composite subtype is in an alias declaration, to get an alternative view of an array object. The rules for determining the index ranges in the view are slightly different from those of type conversions. To start with, the subtype in an alias declaration must have the same base type as that of the object being aliased. This means that the bounds of index ranges of the aliased object are guaranteed to belong to the index subtypes of the alias. If the alias subtype defines index ranges at any given index position, then those index ranges are used for the alias. On the other hand, if index ranges are not defined, then the corresponding index ranges of the aliased object are used for the alias also.

EXAMPLE 3.2 *Alias of a register file signal*

We can declare a register file as follows:

```
type register_array is array (natural range <>) of bit_vector;
signal register_file : register_array(0 to 15)(31 downto 0);
```

We can then declare aliases for

```
alias bigendian_register_file : register_array(open)(0 to 31) is
        register_file;
```

This alias views the register file as an array with the same index range as the original, 0 to 15, since the subtype indication does not specify a top-level index range. Each element, however, is viewed with the index range 0 to 31 specified in the subtype indication.

One common use of aliases is to provide a normalized view of an unconstrained port or parameter so that we can write for loops that iterate over corresponding elements of two potentially different index ranges. In earlier versions of VHDL, using the 'length attribute was sufficient, since any elements of an unconstrained array type had to be constrained with a specific index range. In VHDL-2008, that is no longer the case. We may have to deal with two ports (or two parameters) that have different index ranges for their elements as well as for the top-level arrays. To help us with such situations, VHDL-2008 predefines a new attribute, 'element, that gives the element subtype of an array object, complete with constraints defining the index ranges for the array object. The attribute is also defined for array subtypes, in which case it just gives the element subtype.

EXAMPLE 3.3 *Aliases for normalizing subelements*

We can write a function that locates the first bit difference between two arrays of bit vectors as follows:

```
type bv_vector is array (natural range <>) of bit_vector;

function find_first_difference ( s1, s2 : in bv_vector)
                                    return natural is
  alias s1_norm : bv_vector(0 to s1'length - 1)
                           (0 to s1'element'length - 1) is s1;
  alias s2_norm : bv_vector(0 to s2'length - 1)
                           (0 to s2'element'length - 1) is s2;
  variable count : natural := 0;
begin
  assert s1'length = s2'length and
         s1'element'length = s2'element'length;
  for i in s1_norm'range loop
    for j in s1_norm'element'range loop
```

```
      exit when s1_norm(i)(j) /= s2_norm(i)(j);
      count := count + 1;
    end loop;
  end loop;
  return count;
end function find_first_difference;
```

The two parameters are of an unconstrained type, allowing the function to operate on arrays of various lengths and on arrays with various bit-vector element lengths. The function only requires that, on each call, the two actual parameters have the same shape. In order to deal with the differences, the function declares aliases for the parameters. It views each parameter with an index range starting at 0 and ascending to one less than the length. It views the elements similarly, with an index range starting at 0 and ascending to one less than length of each bit-vector element. The alias declaration uses the **'element** attribute to get the constrained subtype for the actual parameter's elements. Within the function body, the inner for loop also uses the **'element** attribute to get the index range for the elements of the aliases.

VHDL-2008 also predefines the **'subtype** attribute for objects. It provides the subtype of the object, complete with constraints defining index ranges if the object is an array or has any array subelements. Since the subtype is fully constrained, we can use it to declare an object with the same index ranges as the actual associated with an unconstrained or partially constrained formal.

EXAMPLE 3.4   *Swapping variables with unconstrained subelements*

Given the type **bv_array** as defined in Example 3.3, we can declare a procedure to swap two variables of the type:

```
procedure swap_bv_arrays ( a1, a2 : inout bv_array ) is
  variable temp : a1'subtype;
begin
  assert a1'length = a2'length and
         a1'element'length = a2'element'length;
  temp := a1; a1 := a2; a2 := temp;
end procedure swap;
```

Since the type **bv_array** is not fully constrained, we cannot use it as the type of the variable **temp**. Instead, we use the **'subtype** attribute to get a fully constrained subtype with the same shape as **a1**. Once we've verified that **a1** and **a2** are the same shape, we can then swap their values in the usual way using **temp** as the intermediate variable.

Resolved Composite Subtypes

The final place to consider for use of composite subtypes is declaration of resolution functions. To declare a resolution function for signals of a given type, we write a function that takes as a parameter an array with elements of that type and that returns a value of the type. The parameter array type must have an undefined index range, so that signals with different numbers of drivers can be resolved. In earlier versions of VHDL, an unconstrained array type had to have a constrained element subtype. As a consequence, if we wanted to use resolved signals of a composite type, the signal type had to be constrained. We could not, for example, specify a resolved subtype for signals of type bit_vector, and use it for a mixture of 8-bit, 16-bit, and other length signals. There was no way for us to express the resolution function. In VHDL-2008, since we can leave the element subtype of an array unconstrained, we can develop resolved composite subtypes that are unconstrained. The only requirement on the subtype for the resolution function parameter is that it be an array with unconstrained index range. The element subtype can be fully constrained, partially constrained, or unconstrained.

EXAMPLE 3.5   *Resolved unconstrained composite signals*

Suppose, in the interest of simulation performance for a particular application, we want to use signals of type bit_vector, resolved using a bit-wise wired-or operation. The declarations we need are:

```
type bit_vector_vector is array (integer range <>) of bit_vector;

function resolve_vectors ( v : bit_vector_vector )
                         return bit_vector is
  variable result : bit_vector(v'element'range)
           := (others => '0');
begin
  for i in v'range loop
    result := result or v(i);
  end loop;
  return result;
end function resolve_vectors;

subtype resolved_bit_vector is resolve_vectors bit_vector;
```

In the design, if we declare a signal as follows:

```
signal data_bus : resolved_bit_vector(31 downto 0);
```

and drive it with four sources, the resolution function will be passed an array of four 32-bit elements, and will be expected to return a 32-bit result.

3.2   **Resolved Elements**

In earlier versions of VHDL, we could declare resolved subtypes and resolved signals to model signals with multiple sources. This feature is preserved in VHDL-2008. We can associate a *resolution function* with a subtype or signal. The purpose of the resolution function is to determine the resolved value of a signal from the values of the contributing sources. The source values are passed to the function as an array whose element type is the same as that of the signal, and the result type of the function is also that of the signal. Ideally, we declare a signal intended to have only one source with an unresolved subtype, and we declare a signal intended to have multiple sources with a resolved subtype. That way, tools can detect inadvertent connection of multiple sources to signal intended to have only one source.

While this approach works well for scalar subtype and signals, in earlier versions of VHDL it led to problems with array signals. We can illustrate the problem using the types std_ulogic_vector and std_logic_vector, declared in earlier versions as:

```
type std_ulogic is ( ... );
type std_ulogic_vector is array (natural range <>) of std_ulogic;
subtype std_logic is resolved std_ulogic;
type std_logic_vector is array (natural range <>) of std_logic;
```

We would like to be able to use type std_ulogic_vector for signals with only one source per element, and type std_logic_vector for signal with multiple sources per element, for example:

```
signal s1 : std_ulogic_vector(31 downto 0);
signal s2 : std_logic_vector(31 downto 0);
```

However, we could not assign the value of one of these signals to the other, unless we included a type conversion:

```
s1 <= std_ulogic_vector(s2);
```

even though we could assign respective elements:

```
s1(0) <= s2(0);
```

The reason was that for the element types, std_logic was a subtype of std_ulogic, whereas std_logic_vector and std_ulogic_vector were, in earlier versions of the language, two distinct base types declared by distinct type declarations.

A similar problem arose when we connected signals to ports of components. If a component had a port of type std_ulogic_vector, because it had only one source for the port internally, we could not simply connect the port to a signal of type std_ulogic_vector, even if the port was the only source for the signal. Instead, we needed to include a type conversion in the port map:

```
signal s : std_ulogic_vector(0 to 7);
component c is
  port ( p : out std_ulogic_vector(0 to 7); ... );
```

```
end component c;
...

inst : component c
  port map ( std_ulogic_vector(p) => s, ... );
```

While these problems have been part of VHDL since the first version of the standard, devising a way to fix them has proved to be difficult. Nonetheless, an approach has been found and incorporated in VHDL-2008. It involves a way of associating a resolution function with an element type of a composite subtype as part of declaring a new subtype. For example, in VHDL-2008, the type **std_logic_vector** is now defined to be a subtype of **std_ulogic_vector**, declared as:

```
subtype std_logic_vector is (resolved) std_ulogic_vector;
```

The parentheses around the resolution function name, **resolved**, indicates that the resolution function is associated with each element of the array type, rather than with the array type as a whole. Since **std_logic_vector** is now a subtype of **std_ulogic_vector**, not a distinct type, we can freely assign and associate signals and ports of the two types.

The change made in VHDL-2008 is to allow a more general form of *resolution indication* to be included in a subtype indication or signal declaration, rather than just naming a resolution function by itself. The change is backward compatible. If we want to associate a resolution function with an entire subtype, the resolution indication just consists of the resolution function name, as in previous version of VHDL. For example, in the declaration of **std_logic**:

```
subtype std_logic is resolved std_ulogic;
```

The resolution indication is just the resolution function name, **resolved**. In the case of an array whose elements are to be resolved, we write the resolution function name in parentheses, as in the declaration of **std_logic_vector**. We can also resolve the elements of an array type that is itself an array element type. For example, given the following declaration:

```
type unresolved_RAM_content_type is
  array (natural range <>) of std_ulogic_vector;
```

we can declare a subtype with resolved nested elements:

```
subtype RAM_content_type is
  ((resolved)) unresolved_RAM_content_type;
```

The degree of nesting of parentheses indicates how deeply nested in the type structure the resolution function is associated. Two levels indicate that the resolution function is associated with the elements of the elements of the type.

If we have a record type, one of whose elements is to be resolved, we include the element name in the resolution indication. For example, given the following record type with no associated resolution information:

```
type unresolved_status_type is record
  valid : std_ulogic;
  dirty : std_ulogic;
  tag : std_ulogic_vector;
end record unresolved_status_type;
```

we can declare a subtype with a resolved **valid** element as follows:

```
subtype status_resolved_valid is
  (valid wand) unresolved_status_type;
```

We can include resolution functions with multiple elements of the record type by listing the element names and the resolution function associated with each, for example:

```
subtype status_resolved_flags is
  (valid wand, dirty wor) unresolved_status_type;
```

For a record element that is itself of a composite type, we can associate a resolution function with subelements of the record element by writing a parenthesized resolution indication for the element. Thus, to resolve the elements of the **tag** element of the above record type, we would declare a subtype as follows:

```
subtype status_resolved_tag is
  (tag(resolved)) unresolved_status_type;
```

We could combine all of these examples together, resolving all of the scalar subelements, as follows:

```
subtype resolved_status_type is
  (tag(resolved), valid wand, dirty wor) unresolved_status_type;
```

This declaration illustrates that we do not have to write the resolution indications for the record elements in the same order as the declaration of elements in the record types. The record element names in the resolution indication determine the element with which the resolution function is associated.

EXAMPLE 3.6 *Memory system with tristate bus*

We can write a model for a memory system composed of multiple memory devices with tristate data buses. The entity declaration for the memory system is:

```
library ieee; context ieee.ieee_std_context;
entity memory_1Mx8 is
  port ( ce_n, oe_n, we_n : in std_ulogic;
         a : in unsigned(19 downto 0);
         d : inout std_logic_vector(7 downto 0) );
end entity memory_1Mx8;
```

The **d** port is of type **std_logic_vector**, since internally there are multiple sources, one per memory device. The structural architecture is:

```
architecture struct of memory_1Mx8 is
  component memory_256Kx8 is
    port ( ce_n, oe_n, we_n : in std_ulogic;
           a : in unsigned(17 downto 0);
           d : inout std_ulogic_vector(7 downto 0) );
  end component memory_256Kx8;
  signal ce_decoded_n : std_ulogic_vector(3 downto 0);
begin
  with to_x01(a(19 downto 18)) select
    ce_decoded_n <= "1110" when "00",
                    "1101" when "01",
                    "1011" when "10",
                    "0111" when "11",
                    "XXXX" when others;
  chip_gen : for i in 3 downto 0 generate
    chip : component memory_256Kx8
      port map ( ce_n => ce_decoded_n(i),
                 oe_n => oe_n, we_n => we_n,
                 a => a(17 downto 0), d => d );
  end generate chip_gen;
end architecture struct;
```

The **d** port of the component representing the memory devices is of type **std_ulogic_vector**, since each device has only one internal source. Nonetheless, we can connect the **d** port of each instance directly to the **d** port of the memory system entity without type conversion. Had we inadvertently declared the **d** port of the entity to be of type **std_ulogic_vector**, the analyzer would detect the error arising from multiple sources connected to the unresolved elements.

Chapter 4

New and Changed Operations

In VHDL, we model computation by writing expressions that involve application of operations (operators and functions) to operand values. Each operand is of some type, either predefined or user-defined. VHDL defines overloaded versions of operations to perform computation on values of various types. In VHDL-2008, a number of new operations are introduced, and the variety of types to which existing operations can be applied is expanded. We describe the new and changed operations in this chapter.

One point to note is that in earlier versions of VHDL, many of the operations were defined in separate standards. In particular, IEEE Std 1164 specified the package std_logic_1164, which defined the types **std_ulogic**, **std_logic**, **std_ulogic_vector**, and **std_logic_vector** and the operations on those types. Also, IEEE Std 1076.3 specified the packages **numeric_bit** and **numeric_std**, each of which defined the types **unsigned** and **signed** and operations on those types. All of these packages are now included as part of the VHDL-2008 standard. Other changes to the standard packages are described in Chapters 7 and 8.

4.1 Array/Scalar Logical Operations

VHDL provides logical operators, **and**, **or**, **nand**, **nor**, **xor**, and **xnor**, that each operate on a pair of values to produce a result. Earlier versions of VHDL provided predefined and standard overloaded definitions of these operators with the following signatures:

```
[ScalarType, ScalarType return ScalarType]
[ArrayType,  ArrayType  return ArrayType]
```

ScalarType included the types **bit**, **boolean**, and **std_ulogic**, and *ArrayType* included arrays of **bit**, **boolean**, and **std_ulogic** (**std_ulogic_vector**, **std_logic_vector**, **unsigned**, and **signed**). While this was sufficient for many purposes, there were some common modeling problems that required one operand to be an array and the other a scalar, yielding an array result. To meet this requirement, VHDL-2008 provides further overloaded definitions of the logical operators with the following signatures:

```
[ArrayType,        ArrayElementType return ArrayType]
[ArrayElementType, ArrayType        return ArrayType]
```

As before, *ArrayType* includes arrays of **bit**, **boolean**, and **std_ulogic**. *ArrayElementType* is the scalar element type of the other operand. Thus, for example, we can apply an operator such as **and** to a bit_vector and a bit operand. The bit value is ANDed with each element of the array to produce an array result.

EXAMPLE 4.1 *Select logic*

A common problem in coding register read logic is using select bits (each a scalar value) to select among several registers (array values). One possible coding that could be used in earlier versions of VHDL is:

```
genloop : for i in data_bus'range generate
begin
  data_bus(i) <= (a(i) and a_sel) or
                 (b(i) and b_sel) or
                 (c(i) and c_sel);
end generate;
```

Here, a_sel, b_sel and c_sel are the scalar select signals, and a, b, and c are the register values. Note that this coding requires the array indices for the register values and the data bus to be the same. An alternate solution that uses intermediate signals is:

```
signal va_sel, vb_sel, vc_sel :
          std_logic_vector(data_bus'range);
...

va_sel <= (others => a_sel);
vb_sel <= (others => b_sel);
vc_sel <= (others => c_sel);
data_bus <= (a and va_sel) or (b and vb_sel) or (c and vc_sel);
```

Note that this solution does not require the array indices to be the same. The following third alternative is functionally correct if the select signals are mutually exclusive; however, for larger sized "AND-OR" logic, it results in an inefficient hardware implementation known as priority select logic.

```
data_bus <= a when a_sel = '1' else
            b when b_sel = '1' else
            c when c_sel = '1' else
            (others => '0');
```

With the new overloaded definitions introduced in VHDL-2008, these alternatives can be replaced with the simple assignment:

```
data_bus <= (a and a_sel) or (b and b_sel) or (c and c_sel);
```

In each **and** term, the scalar value is applied to each bit of the array value. In effect, the scalar value is replicated in the same manner as the intermediate signal solution, but the statement is much more succinct.

4.2 Array/Scalar Addition Operators

The **numeric_bit** and **numeric_std** packages define overloaded addition ("+") and subtraction ("−") operators for **unsigned** and **signed** operands. Prior to VHDL-2008, if we wanted to code an addition with carry in, we had to convert the carry in to an array value, as follows:

```
signal c_in  : std_logic;
signal a, b  : unsigned(7 downto 0);
signal adder : unsigned(8 downto 0);
...

adder <= ('0' & a) + ('0' & b) + ("" & c_in);
```

Unfortunately, many synthesis tools saw the converted carry in as an additional array value and implemented two adders, where just a single adder with carry in would suffice.

In VHDL-2008, additional overloaded version of the "+" and "−" operators are added to allow the use of a scalar value (such as a **std_logic** value) with an array value. Hence, we can rewrite the above code as follows:

```
adder <= ('0' & a) + ('0' & b) + c_in;
```

It is interesting to note that this same overloading is supported in one vendor's nonstandard synthesis package and results in a single adder. The signatures for the new overloadings are:

```
[ArrayType,        ArrayElementType return ArrayType]
[ArrayElementType, ArrayType        return ArrayType]
```

ArrayType includes **unsigned** and **signed** defined in the **numeric_bit** and **numeric_std** packages. *ArrayElementType* is **bit** (for the **numeric_bit** operators) or **std_ulogic** (for the **numeric_std** operators). The same signatures are also defined for types **ufixed** and **sfixed** defined in the new fixed-point packages (see Section 8.4) and for the type **float** defined in the new floating-point packages (see Section 8.5).

EXAMPLE 4.2 *A conditional incrementer*

The new overloading for the "+" operator allows us to use a scalar control signal as an operand in a conditional incrementer. If the control signal is '0', an **unsigned** value is not incremented; if the control signal is '1', the value is incremented. The declarations and process are:

```
signal inc_en  : std_logic;
signal inc_reg : unsigned(7 downto 0);
...

inc_reg_proc : process (clk) is
```

```
  begin
    if rising_edge(clk) then
      inc_reg <= inc_reg + inc_en;
    end if;
end process inc_reg_proc;
```

Prior VHDL-2008, the conditional incementer would have been coded as:

```
inc_reg_proc : process (clk) is
begin
  if rising_edge(clk) then
    if inc_en = '1' then
      inc_reg <= inc_reg + 1;
    end if;
  end if;
end process inc_reg_proc;
```

The use of the integer value 1 as an operand would have implied an adder, rather than just an incrementer.

4.3 Logical Reduction Operators

In earlier versions of VHDL, the logical operators **and**, **or**, **nand**, **nor**, **xor**, and **xnor** were defined only as binary operators; that is, they each operated on two operands. The operands could be **bit** or **boolean** values, or they could be arrays of **bit** or **boolean** elements. In the case of array operands, the logical operator is applied to corresponding array elements to produce an array result. In some models, we need to apply a logical operator to all of the elements of an array to produce a single scalar result. To do this in earlier versions of VHDL, we had to write a loop to apply the operator to the elements.

VHDL-2008 extends the definition of logical operators to allow them to be used as unary operators. Each such *logical reduction operator* is applied to a single operand that is an array of **bit** or **boolean** elements and produces a **bit** or **boolean** result, respectively. The **std_logic_1164** package also defines overloaded logical reduction operators for **std_ulogic_vector** operands. Thus, the signature of each logical reduction operator is:

[ArrayType **return** *ArrayElementType]*

The reduction **and**, **or**, and **xor** operators form the logical AND, OR, and exclusive OR, respectively of the array elements. Thus:

```
and "0110" = '0' and '1' and '1' and '0' = '0'

or  "0110" = '0' or  '1' or  '1' or  '0' = '1'

xor "0110" = '0' xor '1' xor '1' xor '0' = '0'
```

In each case, if the array has only one element, the result is the value of that element. If the array is a null array (that is, it has no elements), the result of the **and** operator is '1', and the result of the **or** and **xor** operators is '0'.

The reduction **nand**, **nor**, and **xnor** operators are the negation of the reduction **and**, **or**, and **xor** operators, respectively. Thus:

nand "0110" = **not** ('0' **and** '1' **and** '1' **and** '0') = **not** '0' = '1'

nor "0110" = **not** ('0' **or** '1' **or** '1' **or** '0') = **not** '1' = '0'

xnor "0110" = **not** ('0' **xor** '1' **xor** '1' **xor** '0') = **not** '0' = '1'

In each case, application to a single-element array produces the negation of the element value. Application of **nand** to a null array produces '0', and application of **nor** or **xnor** to a null array produces '1'.

The logical reduction operators have the same precedence as the unary **not** and **abs** operators. In the absence of parentheses, they are evaluated before binary operators. So the expression:

and A **or** B

involves applying the reduction **and** operator to A, then applying the binary **or** operator to the result and B. In some cases, we need to include parentheses to make an expression legal. For example, the expression:

and not X

is not legal without parentheses, since we cannot chain unary operators. Instead, we must write the expression as:

and (**not** X)

EXAMPLE 4.3   *Parity of a vector value*

Without reduction operators, calculating parity requires the following:

```
parity <= data(7) xor data(6) xor data(5) xor data(4) xor
          data(3) xor data(2) xor data(1) xor data(0);
```

With reduction operators, calculating parity becomes

```
parity <= xor Data;
```

Since reduction operators have higher precedence than binary logical operators, the following two asignments produce the same value:

```
parity_error1 <= (xor data) and received_parity;
parity_error2 <= xor data and received_parity;
```

However, for readability, parentheses are recommended.

4.4 Condition Operator

VHDL provides numerous language constructs that use a condition to control what actions are performed. A condition is an expression that produces a **boolean** result, for example, through application of relational and logical operators. In earlier versions of VHDL, the fact that a condition had to produce a **boolean** value was a source of inconvenience, particularly in models that used **bit** or **std_ulogic** values for control signals. We would typically write a condition using such a control signal as:

```
if control_sig = '1' then ...
```

VHDL-2008 provides two new language features that allow us to treat an expression producing a **bit** or **std_ulogic** value as a condition. The first of these features is a *condition operator*, "??", that converts from a **bit** or **std_ulogic** value to a **boolean** value. For **bit**, "??" converts '1' to true and '0' to false. For **std_ulogic**, "??" converts both '1' and 'H' to true and all other values to false. (We can also overload the operator for other user-defined types.) Thus, we could rewrite the if-statement condition shown above as:

```
if ?? control_sig then ...
```

Normally, we would not apply the condition operator explicitly like this, as the second of the new features involves implicit application of the operator in conditions. The operator is implicitly applied in a condition when the expression could otherwise not be interpreted as producing a **boolean** result and there is a unique interpretation using the condition operator that does produce a **boolean** result. These potential interpretations require use of the rules for resolving overloaded operators to determine the type of the expression. Note that an ambiguous **boolean** expression is considered a **boolean** interpretation and still results in an error.

As an example, assuming **control_sig** is a **std_ulogic** signal, we would rewrite the if-statement condition shown above as:

```
if control_sig then ...
```

The condition operation is implicitly applied, since that is the one and only way of getting a **boolean** result from the expression.

The places where the condition operator is considered for application are:

- after **until** in a wait statement

- after **assert** in an assertion statement

- after **while** in a while loop

- after **if** or **elsif** in an if statement

- after **when** in a next statement or exit statement

- after **when** in a conditional signal or variable assignment statement

- after **if** or **elsif** in an if-generate statement

- in a Boolean expression in a PSL declaration or a PSL directive

This list includes all of the cases where an expression of type **boolean** was required in earlier versions of VHDL.

EXAMPLE 4.4   *Std_logic control conditions*

With the condition operator implicitly applied, we can write the following in VHDL-2008:

```
signal cs1, ncs2, cs3 : std_logic;
...

if cs1 and not cs2 and cs3 then
  ...
```

Backward compatibility is maintained, so we can still write:

```
if cs1 = '1' and ncs2 = '0' and cs3 = '1' then
  ...
```

Note, however, that we cannot write a condition that mixes **std_ulogic** and **boolean** operands for a logical operator, such as:

```
if cs1 and cs3 and ncs2 = '0' then   -- illegal
  ...
```

The "??" operator is only implicitly applied to the entire condition. There is no overloading for **and** that has a **std_logic** left operand and a **boolean** right operand.

4.5   Matching Relational Operators

VHDL provides ordinary relational operators ("=", "/=", "<", "<=", ">", and ">=") that return a result of type **boolean**. We can use the result as a condition to control what actions are performed in a model. However, if we want to use the result to assign to a signal of type **bit** or **std_ulogic**, we have to resort to a form such as:

```
control_sig <= '1' when X = Y else '0';
```

VHDL-2008 has a new set of predefined *matching relational operators* ("?=", "?/=", "?<", "?<=", "?>", and "?>=") that return **bit** or **std_ulogic** results. This allows us to rewrite the assignment as:

```
control_sig <= X ?= Y;
```

The matching relational operators are predefined with the following signatures for scalar operands:

```
[ScalarType, ScalarType return ScalarType]
```

ScalarType is one of **bit** or **std_ulogic**. For **bit** operands, the results are the same as the ordinary relational operators, except that the matching relational versions return '0' or '1' instead of false or true. For **std_ulogic** operands, the result values are shown in Tables 4.1, 4.2, and 4.3. VHDL-2008 lists the result values for the "?=" and "?<" operators, and then defines the results for the remaining operators using the **not** and **or** operators for **std_ulogic**. Note that for "?<", "?<=", "?>", and "?>=", an operand value of '–' produces an assertion-violation error, so the seemingly anomalous results shown in Tables 4.2 and 4.3 are not a concern. They are defined for completeness in case we chose to ignore assertion violations during simulation.

In addition, the "?=" and "?/=" operators are predefined with the signature:

[*ArrayType*,   *ArrayType*   **return** *ArrayElementType*]

where ***ArrayType*** is any one dimensional array of **bit** or **std_ulogic** elements. The array operands must be of the same length. The array "?=" operator applies the scalar "?=" operator to corresponding elements of the arrays, and then forms the logical AND of the resulting values. The array "?/=" operator does the same, but negates the final result.

We can also overload all of these operators. In particular, the "?<", "?<=", "?>", and "?>=" operators are not predefined for arrays of **bit** or **std_ulogic** elements; instead they are overloaded in the appropriate numeric packages. They are overloaded for **signed** and **unsigned** in **numeric_bit** and **numeric_std**; for **ufixed** and **sfixed** in **fixed_generic_pkg**; for **float** in **float_generic_pkg**; for **bit_vector** in **numeric_bit_unsigned**; and for **std_ulogic_vector** in **numeric_std_unsigned**. (See Chapter 8 for more details about the standard packages.) Thus, we must include a use clause for the appropriate package if we want to apply the operators to **bit_vector** or **std_logic** vector operands.

TABLE 4.1   *Result values for the "?=" and "?/=" operators on* **std_ulogic** *operands*

?=	Right								
Left	'U'	'X'	'0'	'1'	'Z'	'W'	'L'	'H'	'-'
'U'	'U'	'U'	'U'	'U'	'U'	'U'	'U'	'U'	'1'
'X'	'U'	'X'	'X'	'X'	'X'	'X'	'X'	'X'	'1'
'0'	'U'	'X'	'1'	'0'	'X'	'X'	'1'	'0'	'1'
'1'	'U'	'X'	'0'	'1'	'X'	'X'	'0'	'1'	'1'
'Z'	'U'	'X'	'X'	'X'	'X'	'X'	'X'	'X'	'1'
'W'	'U'	'X'	'X'	'X'	'X'	'X'	'X'	'X'	'1'
'L'	'U'	'X'	'1'	'0'	'X'	'X'	'1'	'0'	'1'
'H'	'U'	'X'	'0'	'1'	'X'	'X'	'0'	'1'	'1'
'-'	'U'	'1'	'1'	'1'	'1'	'1'	'1'	'1'	'1'

?/=	Right								
Left	'U'	'X'	'0'	'1'	'Z'	'W'	'L'	'H'	'-'
'U'	'U'	'U'	'U'	'U'	'U'	'U'	'U'	'U'	'0'
'X'	'U'	'X'	'X'	'X'	'X'	'X'	'X'	'X'	'0'
'0'	'U'	'X'	'0'	'1'	'X'	'X'	'0'	'1'	'0'
'1'	'U'	'X'	'1'	'0'	'X'	'X'	'1'	'0'	'0'
'Z'	'U'	'X'	'X'	'X'	'X'	'X'	'X'	'X'	'0'
'W'	'U'	'X'	'X'	'X'	'X'	'X'	'X'	'X'	'0'
'L'	'U'	'X'	'0'	'1'	'X'	'X'	'0'	'1'	'0'
'H'	'U'	'X'	'1'	'0'	'X'	'X'	'1'	'0'	'0'
'-'	'U'	'0'	'0'	'0'	'0'	'0'	'0'	'0'	'0'

TABLE 4.2 Result values for the "?<" and "?<=" operators on std_ulogic operands

?<= Left \ Right	'U'	'X'	'0'	'1'	'Z'	'W'	'L'	'H'	'-'
'U'	'U'	'U'	'U'	'U'	'U'	'U'	'U'	'U'	'X'
'X'	'U'	'X'	'X'	'X'	'X'	'X'	'X'	'X'	'X'
'0'	'U'	'X'	'1'	'1'	'X'	'X'	'1'	'1'	'X'
'1'	'U'	'X'	'0'	'1'	'X'	'X'	'0'	'1'	'X'
'Z'	'U'	'X'	'X'	'X'	'X'	'X'	'X'	'X'	'X'
'W'	'U'	'X'	'X'	'X'	'X'	'X'	'X'	'X'	'X'
'L'	'U'	'X'	'1'	'1'	'X'	'X'	'1'	'1'	'X'
'H'	'U'	'X'	'0'	'1'	'X'	'X'	'0'	'1'	'X'
'-'	'X'	'X'	'X'	'X'	'X'	'X'	'X'	'X'	'X'

?< Left \ Right	'U'	'X'	'0'	'1'	'Z'	'W'	'L'	'H'	'-'
'U'	'U'	'U'	'U'	'U'	'U'	'U'	'U'	'U'	'X'
'X'	'U'	'X'	'X'	'X'	'X'	'X'	'X'	'X'	'X'
'0'	'U'	'X'	'0'	'1'	'X'	'X'	'0'	'1'	'X'
'1'	'U'	'X'	'0'	'0'	'X'	'X'	'0'	'0'	'X'
'Z'	'U'	'X'	'X'	'X'	'X'	'X'	'X'	'X'	'X'
'W'	'U'	'X'	'X'	'X'	'X'	'X'	'X'	'X'	'X'
'L'	'U'	'X'	'0'	'1'	'X'	'X'	'0'	'1'	'X'
'H'	'U'	'X'	'0'	'0'	'X'	'X'	'0'	'0'	'X'
'-'	'X'	'X'	'X'	'X'	'X'	'X'	'X'	'X'	'X'

TABLE 4.3   *Result values for the "?>" and "?>=" operators on std_ulogic operands*

?>	Right								
Left	'U'	'X'	'0'	'1'	'Z'	'W'	'L'	'H'	'-'
'U'	'U'	'U'	'U'	'U'	'U'	'U'	'U'	'U'	'0'
'X'	'U'	'X'	'X'	'X'	'X'	'X'	'X'	'X'	'0'
'0'	'U'	'X'	'0'	'0'	'X'	'X'	'0'	'0'	'0'
'1'	'U'	'X'	'1'	'0'	'X'	'X'	'1'	'0'	'0'
'Z'	'U'	'X'	'X'	'X'	'X'	'X'	'X'	'X'	'0'
'W'	'U'	'X'	'X'	'X'	'X'	'X'	'X'	'X'	'0'
'L'	'U'	'X'	'0'	'0'	'X'	'X'	'0'	'0'	'0'
'H'	'U'	'X'	'1'	'0'	'X'	'X'	'1'	'0'	'0'
'-'	'0'	'0'	'0'	'0'	'0'	'0'	'0'	'0'	'0'

?>=	Right								
Left	'U'	'X'	'0'	'1'	'Z'	'W'	'L'	'H'	'-'
'U'	'U'	'U'	'U'	'U'	'U'	'U'	'U'	'U'	'X'
'X'	'U'	'X'	'X'	'X'	'X'	'X'	'X'	'X'	'X'
'0'	'U'	'X'	'1'	'0'	'X'	'X'	'1'	'0'	'X'
'1'	'U'	'X'	'1'	'1'	'X'	'X'	'1'	'1'	'X'
'Z'	'U'	'X'	'X'	'X'	'X'	'X'	'X'	'X'	'X'
'W'	'U'	'X'	'X'	'X'	'X'	'X'	'X'	'X'	'X'
'L'	'U'	'X'	'1'	'0'	'X'	'X'	'1'	'0'	'X'
'H'	'U'	'X'	'1'	'1'	'X'	'X'	'1'	'1'	'X'
'-'	'X'	'X'	'X'	'X'	'X'	'X'	'X'	'X'	'X'

EXAMPLE 4.5   *Assignment of a condition result for a select signal*

We can write a Boolean equation for a **std_ulogic** select signal that includes chip-select control signals and an address signal. In earlier versions of VHDL, we had to write an assignment in the following form, including a condition of type **boolean**:

```
dev_sel1 <= '1' when cs1 = '1' and
                    ncs2 = '0' and addr = X"A5" else '0';
```

In VHDL-2008, we can use the "?=" operator, which returns a **std_ulogic** result. We can combine that result with the **std_ulogic** control signals to produce a **std_ulogic** form of the Boolean equation:

```
dev_sel1 <= cs1 and not ncs2 and addr ?= X"A5";
```

We can also use this form of expression in a condition, since the condition operator, "??" (see Section 4.4), is implicitly applied:

```
if cs1 and not ncs2 and addr ?= X"A5" then
    . . .
```

or similarly:

```
if cs1 and ncs2 ?= '0' and addr ?= X"A5" then
    . . .
```

Note that, in a condition, we still have backward compatibility. The following forms are still valid:

```
if cs1 = '1' and ncs2 = '0' and addr = X"A5" then
    . . .
```

and

```
if (cs1 and not ncs2) = '1' and addr = X"A5" then
    . . .
```

4.6   Maximum and Minimum

If we want to find the larger or the smaller of two values, we can write an if statement, such as:

```
if A > B then
  greater := A;
else
  greater := B;
end if;
```

However, in some cases, it would be more convenient to select the larger or smaller value as part of an expression. VHDL-2008 allows us to do so using new predefined **maximum** and **minimum** functions, with the following signatures:

```
[ScalarType, ScalarType return ScalarType]
[DiscreteArrayType,  DiscreteArrayType
    return DiscreteArrayType]
```

Here, *ScalarType* is any scalar type, and *DiscreteArrayType* is any discrete array type (that is, an array type whose elements are of an integer or enumeration type). These are the types for which the operator "<" is predefined, and the results of the **maximum** and **minimum** functions are defined in terms of the "<" operator applied to the operands. For example:

```
maximum(3, 20) = 20        minimum(3, 20) = 3

maximum('a', 'z') = 'z'    minimum('a', 'z') = 'a'
```

Note that for array types, the "<" operator uses dictionary ordering to compare operands. The two arrays do not have to be of the same length. Corresponding elements are compared, from left to right, until a pair with differing values is encountered (in which case the lesser array is the one containing the lesser element of the pair) or until the end of one array is reached (in which case the lesser array is the shorter of the two). Since the **maximum** and **minimum** functions are defined in terms of the "<" operator, they also use dictionary ordering for arrays, for example:

```
maximum(bit_vector'("101"), bit_vector'("100100")) = "101"

minimum(bit_vector'("101"), bit_vector'("100100")) = "100100"
```

As these examples show, for vectors representing binary-coded numeric values, the predefined **maximum** and **minimum** functions are not consistent with numerical ordering. (The same argument also applies to the predefined relational operators.) For this reason, the standard numeric packages define overloaded versions of **maximum** and **minimum** (and the relational operators) that do give results consistent with numerical ordering.

EXAMPLE 4.6   *Maximum function used in a declaration*

One use of the **maximum** and **minimum** is in expressions in declarations. For example, in the following function for saturating addition of unsigned numeric values, we use the **maximum** function to determine the longer of the two operand values, and then declare the result variable to be of that size.

```
function saturating_add (A, B : unsigned) return unsigned is
  constant size : natural := maximum(A'length, B'length);
  variable result : unsigned(size - 1 downto 0);
  variable c_out  : std_ulogic;
begin
```

```
      (c_out, result) := ('0' & A) + ('0' & B);
      if c_out then
        result := (others => '1');
      end if;
      return result;
    end function saturating_add;
```

VHDL-2008 also predefines the **maximum** and **minimum** functions as reduction operations on array values. The signature is:

[*ArrayType* **return** *ArrayElementType*]

Here, *ArrayType* is an array of any scalar element type (not just a discrete type), and *ArrayElementType* is the element type. The **maximum** function of this form returns the largest element in the array, and the **minimum** function returns the smallest element in the array. Again, the comparisons are performed using the predefined "<" operator for the element type. Thus,

```
maximum(string'("WYZ")) = 'Z'   minimum(string'("WXYZ")) = 'W'

maximum(time_vector'(10 ns, 50 ns, 20 ns)) = 50 ns

minimum(time_vector'(10 ns, 50 ns, 20 ns)) = 10 ns
```

For a null array (one with no elements), the **maximum** function returns the smallest value of the element type, and the **minimum** functions returns the largest value of the element type.

4.7   Mod and Rem for Physical Types

Prior to VHDL-2008, the arithmetic operators "+", "−", "*" and "/" were predefined for physical types, including type **time**, but the **mod** and **rem** operators were not predefined. VHDL-2008 adds predefined **mod** and **rem** functions for these types, with the following signature.:

[*PhysicalType*, *PhysicalType* **return** *PhysicalType*]

For example, using type **time**:

```
   5 ns  rem    3 ns =    2 ns
   5 ns  mod    3 ns =    2 ns
 (-5 ns) rem    3 ns =   -2 ns
 (-5 ns) mod    3 ns =    1 ns
   1 ns  mod 300 ps  = 100 ps
 (-1 ns) mod 300 ps  = 200 ps
```

EXAMPLE 4.7   *Generating a periodic waveform*

We can use the **mod** operator to simplify generation of a periodic waveform. For example, the following process creates a triangle wave on the real signal triangle_wave. T_period_wave defines the period of the output wave, t_offset defines the offset within the triangle wave, and t_period_sample defines how many points are in the waveform.

```
signal triangle_wave : real;
...


wave_proc : process is
  variable phase : time;
begin
  phase := (now + t_offset) mod t_period_wave;
  if phase <= t_period_wave/2 then
    triangle_wave <= 4.0 * real(phase / t_period_wave) - 1.0;
  else
    triangle_wave <= 3.0 - 4.0 * real(phase / t_period_wave);
  end if;
  wait for tperiod_sample;
end process wave_proc;
```

4.8   Shift Operations

Prior to VHDL-2008, the shift operations (**rol**, **ror**, **sll**, **srl**, **sla**, and **sra**) were predefined only for arrays of **bit** and **boolean** elements. The operations can take a positive shift count, in which case they rotate or shift in the direction suggested by the operator name. They can also take a negative shift count, in which case they rotate or shift in the opposite direction. The rotate and logical-shift operations have the expected meanings. The arithmetic-shift operators also have the expected meaning when shifting right; that is, they replicate the leftmost bit. However, when shifting left, they replicate the rightmost bit, treating it as a sign bit. This seems anomalous to many designers, so the operation is rarely (if ever) used.

The numeric_bit and numeric_std packages used in earlier versions of VHDL defined overloaded versions of the **rol**, **ror**, **sll**, and **srl** operators with similar behavior to that of the predefined operators on bit_vector values. The packages did not, however, overload **sla** and **sra**, preferring instead to define shift_left and shift_right functions that perform logical shifts on **unsigned** values and arithmetic shifts on **signed** values. A shift left on a **signed** value fills the vacated positions with '0', rather than replicating the rightmost bit. This is generally more appropriate for arithmetic circuits.

VHDL-2008 extends the definitions of shift operations to include **sla** and **sra** in numeric_bit and numeric_std. It also defines all of the shift operations in the new arithmetic packages numeric_bit_unsigned, numeric_std_unsigned, and in the fixed-point packages (see Chapter 8). In all cases, the overloaded **sla** and **sra** operators on signed values have the same numeric behavior as the shift_left and shift_right functions.

4.9   Strength Reduction and 'X' Detection

Prior to VHDL-2008, the strength reduction and 'X' detection functions were not uniformly implemented throughout the packages based on the **std_ulogic** type. The package **std_logic_1164** defined the detection function **is_X** and the strength reduction functions **to_X01**, **to_X01Z**, and **to_UX01** for scalar and vector types. The **numeric_std** package, however, did not define these functions for **unsigned** or **signed**. Instead, we had to convert values of those types to **std_logic_vector** in order to use the functions. The package did, however, define the function **to_01** that maps non-logic values to a value of our choice (the default being '0').

In VHDL-2008, the inconsistency is rectified, and the functions are also defined in the new fixed-point and floating-point packages based on the **std_ulogic** type (see Chapter 8). To summarize, the following functions are defined in the packages:

```
is_X    [AType return boolean]
to_X01  [AType return AType]
to_X01Z [AType return AType]
to_UX01 [AType return AType]
```

In **std_logic_1164**, *AType* covers **std_ulogic**, **std_logic**, **std_ulogic_vector** and **std_logic_vector**; in **numeric_std**, *AType* covers **unsigned** and **signed**; in the fixed-point packages, *AType* covers **ufixed** and **sfixed**; and in the floating-point packages, *AType* covers **float** and its subtypes.

In addition, the function **to_01** is defined with the following signature:

```
to_01 [AType, std_ulogic return AType]
```

The second parameter is the value to which non-logic values such as 'X' are mapped. This function is defined in packages **numeric_std**, **numeric_std_unsigned**, the fixed-point packages, and the floating-point packages, with the same types for *AType* as the other strength-reduction functions.

EXAMPLE 4.8   *Strength reduction and 'X' detection in models*

We can use the **to_X01** function in behavioral models and ASIC or FPGA input cells to promote a resistive strength to a driving level as follows:

```
ncs_x01 <= to_X01(ncs);
```

We can use the **is_X** function to detect 'X' values in behavioral models and RTL code, for example, in the input to a state machine:

```
assert not is_X(ncs) report "ncs is X" severity error;
```

Chapter 5

New and Changed Statements

VHDL provides various forms of statements for modeling the behavior of hardware and testbenches. Sequential statements are used to express algorithms within processes and subprograms, where there is just one thread of control. Concurrent statements, on the other hand, express multi-threaded control. They are also used to represent structural decomposition of a design into concurrently operating subsystems.

In this chapter, we look at the enhancements to the statement repertoire in VHDL-2008. We start with changes to assignment statements, which include new sequential forms that mirror conditional and selected concurrent assignments. Next, we look at changes to case statements that allow matching of standard-logic values with don't care elements. Finally, we look at extensions to if-generate statements that allow multiple conditions to be checked, and a new case-generate statement.

5.1 Conditional and Selected Assignments

In earlier versions of VHDL, sequential and concurrent signal assignment statements had different syntactic forms. Sequential signal assignments, appearing in processes and subprograms, could only take the simple form of a target signal on the left-hand side and a list of one or more values and delays on the right-hand side. Concurrent signal assignments, appearing in architectures, could take this simple form, but could also take conditional and selected forms. While we could embed a sequential assignment in an if statement or a case statement, the differences between the sequential and concurrent contexts was a cause for confusion among designers.

In this section, we describe the way VHDL-2008 extends assignments. This includes allowing conditional and selected forms of signal assignments in processes and subprograms, providing for a signal to be forced by a conditional or selected assignment, and providing selected and conditional variable assignments.

5.1.1 Sequential Signal Assignments

VHDL-2008 extends the allowed forms of signal assignments to be consistent between the sequential and concurrent contexts. Within a process or subprogram, we can write conditional and selected signal assignments in the same form as those in architecture bodies. The effect is equivalent to writing simple signal assignments within if statements or case statements, but the notation is more succinct.

EXAMPLE 5.1   *Register process using a conditional assignment*

A process representing a register with synchronous reset can be written using a conditional signal assignment as follows:

```
reg : process (clk) is
begin
  if rising_edge(clk) then
    q <= (others => '0') when reset else d;
  end if;
end process reg;
```

The conditional assignment in the process is equivalent to the if statement:

```
if reset then
  q <= (others => '0')
else
  q <= d;
end if;
```

EXAMPLE 5.2   *Next-state process for a finite-state machine*

Use of selected assignments simplifies description of the next-state logic of a finite-state machine, as is shown by the following process outline:

```
next_state_logic : process (all) is
begin
  with current_state select
    idle =>
      next_state <= pending1 when request and busy      else
                    active1  when request and not busy else
                    idle;
    pending1 =>
      ...
    ...
  end case;
end process next_state_logic;
```

EXAMPLE 5.3   *Combined multiplexer and register using a selected assignment*

We can model a register with a multiplexer at its input in a single process as follows:

```
mux_reg : process (clk) is
begin
  if rising_edge(clk) then
    with d_sel select
```

```
        q <= source0 when "00",
             source1 when "01",
             source2 when "10",
             source3 when "11";
    end if;
end process mux_reg;
```

The selected assignment in the process is equivalent to the case statement:

```
case d_sel is
  when "00" =>
    q <= source0;
  when "01"
    q <= source1;
  when "10"
    q <= source2;
  when "11"
    q <= source3;
end case;
```

When we write a conditional or selected signal assignment in a sequential context, we can include delays and multiple waveform values, just as we do in concurrent contexts. For example, in a stimulus-generator process, we could write the assignment:

```
req <= '1', '0' after T_fixed when fixed_delay_mode else
       '1', '0' after next_random_delay(ran_seed);
```

If we need to include an inertial or transport delay specification in a sequential assignment, we write it in the same way as in a concurrent assignment. For example, a sequential conditional assignment using transport delay could be written as:

```
wire_out <= transport
  wire_in after T_wire_delay when delay_mode = fixed else
  wire_in after delay_lookup("wire_out");
```

Likewise, a sequential conditional assignment using inertial delay could be written as:

```
with speed_grade select
  z <= reject Tpr inertial
    result after Tpd_std when std_grade,
    result after Tpd_fast when fast_grade,
    result after Tpd_redhot when redhot_grade;
```

We can also use the reserved word **unaffected** in a sequential signal assignment to represent no assignment to the target signal, for example:

```
with dut_state select
  dut_req <= '1' when ready,
            '0' when ack,
            unaffected when others;
```

A related change is that we can use the reserved word **unaffected** in a simple sequential signal assignment. This was previously illegal. In VHDL-2008, we can write the following statements within a process or subprogram:

```
if dut_busy then
  collision_count := collision_count + 1;
  dut_req <= unaffected;
else
  accepted_count := accepted_count + 1;
  dut_req <= '1';
end if;
```

The assignment using **unaffected** is the same as doing nothing (using a null statement), but the design intent is explicitly documented. It is clear that we did not inadvertently omit an assignment.

One aspect of concurrent signal assignments that we cannot include in a sequential assignment is the reserved word **guarded**. The effect of including **guarded** in a concurrent assignment is to cause the target signal to be disconnected when a **guard** signal is true, and to reconnect when the **guard** signal becomes false. Reconnection is done by executing the concurrent assignment. Thus, the **guard** signal has some external control over when the concurrent assignment is executed. This would not be appropriate for a sequential assignment, which should only be executed when control reaches it within the enclosing process or subprogram.

5.1.2 Forcing Assignments

In Section 2.2, we described the new features in VHDL-2008 for forcing and releasing signals. Force and release assignments are both forms of sequential signal assignment statements. VHDL-2008 also allows us to write forcing assignments in the form of conditional and selected assignments within processes and subprograms. A conditional forcing assignment has the form

```
signal_name <= force mode
  value when condition else
  ...
```

and a selected forcing assignment has the form

```
with expression select
  signal_name <= force mode
    value when choices,
    ...
```

The *mode* is optional, and can be either **in** or **out** to specify forcing of the effective value or the driving value of the target signal, respectively, as described in Section 2.2. The effect of these statements is to allow us to choose the value to force onto the target, depending on a number of conditions or on the value of an expression. They provide a more succinct way of writing the choice than embedding a number of simple forcing assignments in an if statement or case statement.

EXAMPLE 5.4   *Conditional forcing assignment*

A conditional forcing assignment can be used to choose between a randomly generated stimulus value or a directed-test stimulus value in a loop that applies successive tests. The stimulus value is used to force the effective value of a bidirectional port of a design under test. The code in the testbench is:

```
alias dut_d_bus is
  <<signal dut.d_bus:std_logic_vector(15 downto 0)>>;
...

for test_count in 1 to num_tests loop
  dut_d_bus <= force in
    next_random_stim(dut_d_bus'length)
      when test_mode = random else
    directed_stim(test_count);
  wait for test_interval;
end loop;
```

5.1.3   Variable Assignments

One of the reasons for providing sequential forms of conditional and selected signal assignments in VHDL-2008 is to provide consistency with concurrent signal assignments. In the further interest of consistency, VHDL-2008 also provides conditional and selected forms of variable assignment statement for use in processes and subprograms. A conditional variable assignment has the form

```
variable_name := value when condition else
              ...
```

and a selected variable assignment has the form

```
with expression select
  variable_name := value when choices,
              ...
```

EXAMPLE 5.5   *Conditional assignment for an intermediate variable*

A variable can be used for an intermediate value in a synthesizable process. No actual storage is implied for the variable, provided it is updated on all execution paths through the process. A conditional variable assignment allows us to make assignments to variables in the same succinct form that we can use for signals. Thus, in the process:

```
arith_unit : process (all) is
  variable tmp : operand_type;
begin
  tmp := a - b when mode else a + b;
  new_result <= result + scale * tmp;
end process arith_unit;
```

the assignment to the variable **tmp** is equivalent to the statements:

```
if mode then
  tmp := a - b;
else
  tmp := a + b;
end if;
```

EXAMPLE 5.6   *Selected assignment for combined multiplexer and register*

We can use a selected variable assignment for an intermediate variable representing a multiplexer at the input to a register. The process is:

```
mux_reg : process (clk) is
  variable mux : data_type;
begin
  if rising_edge(clk) then
    with mux_sel select
      mux := in0 when "00",
             in1 when "01",
             in2 when "10",
             in3 when "11",
             (others => 'X') when others;
    reg_out <= (others => '0') when reset else mux;
  end if;
end process mux_reg;
```

5.2 Matching Case Statements

A case statement in VHDL allows us to perform alternative actions depending on the value of an expression. We write choice values in each alternative, immediately preceding the "=>" symbol. The choices are compared for exact equality with the expression value to select an alternative. If the type of the case expression and choices is a vector of std_logic values, the exact comparison is not always what we want. In particular, we would like to be able to include don't care elements ('–') in the choices to indicate that we don't care about some elements of the case expression when selecting an alternative.

VHDL-2008 provides a new form of case statement, called a *matching case statement*, that uses the predefined "?=" operator described in Section 4.5 to compare choice values with the expression value. We include a question mark symbol after the keyword **case**, as follows:

```
case? expression is
  ...
end case?;
```

The most common use of a matching case statement is with an expression of a vector type whose elements are **std_ulogic** or **std_logic** values. That includes the standard types **std_ulogic_vector**, **std_logic_vector**, **unsigned**, **signed**, and so on. It also includes vector types that we might define. With a case expression of such a type, we can write choice values that include '–' elements to specify don't care matching.

EXAMPLE 5.7 *Priority arbiter using don't care matching*

Suppose we have vectors of request and grant signals, declared as follows:

```
signal request, grant : std_logic_vector(0 to 3);
```

We can use a matching case statement in a priority arbiter, with request 0 having highest priority:

```
case? request is
  when "1---" => grant <= "1000";
  when "01--" => grant <= "0100";
  when "001-" => grant <= "0010";
  when "0001" => grant <= "0001";
  when others => grant <= "0000";
end case?;
```

Each choice is compared with the case expression using the predefined "?=" operator. Thus, the first choice matches values "1000", "1001", "100X", "H000", and so on, and similarly for the remaining choices. This is a much more succinct way of describing the arbiter than using an ordinary case statement. Moreover, unlike a sequence of tests in an if statement, it does not imply chained decision logic.

When we use a matching case statement with a vector-type expression, the value of the expression must not include any '–' elements. (This is different from the choice values, which can include '–' elements.) The reason is that an expression value with a '–' element would match multiple choice values, making selection of an alternative ambiguous. Normally, this rule is not a problem, since we don't usually assign '–' values to signals or variables. They usually just occur in literal values for comparison and in testbench assertions.

In an ordinary case statement, we need to include choices for all possible values of the case expression. A related rule applies in a matching case statement. Each possible value of the case expression, except those that include any '–' elements, must be represented by exactly one choice. By "represented," we mean that comparison of the choice and the expression value using the "?=" operator yields '1'. Hence, our choice values would generally just include '0', '1', and '–' elements, matching with '0', 'L', '1', 'H' elements in the case expression value. We could also include 'L' and 'H' elements in a choice. However, we would not include 'U', 'X', 'W', or 'Z' choice elements, since they only ever produce 'U' or 'X' results, and so never match. As with an ordinary case statement, we can include an **others** choice to represent expression values not otherwise represented. Unlike an ordinary case statement, a choice can represent multiple expression values if it contains a '–' element.

We mentioned that a vector type including **std_ulogic** or **std_logic** values is the most common type for a matching case statement. Less commonly, we can write an expression of type **std_ulogic**, **std_logic**, **bit**, or a vector of **bit** elements. These are the other types for which the "?=" operator is predefined. For **std_ulogic** or **std_ulogic** expressions, the choice values would typically be either '0' (matching an expression value of '0' or 'L') or '1' (matching an expression value of '1' or 'H'). We would not write a choice of '–', since that would match all expression values, preventing us from selecting distinct alternatives. For case expressions of type **bit** or a vector of **bit** elements, a matching case statement has exactly the same behavior as an ordinary case statement. VHDL-2008 allows matching case statements of this form to allow synthesizable models to be written uniformly regardless of whether **bit** or **std_logic** data types are used.

5.2.1 Matching Selected Assignments

Selected assignments in VHDL are shorthand notations for assignments within case statements. This applies to concurrent selected signal assignments in earlier versions of VHDL, as well as to sequential selected signal and variable assignments in VHDL-2008. In all of these forms of selected assignment, we can include a "?" symbol after the **select** keyword to indicate that the implied case statement is a matching case statement instead of an ordinary case statement. The rules covering the type of the case expression and the way in which choices are matched then apply to the selected assignment.

EXAMPLE 5.8   *Priority arbiter using matching selected assignment*

We can rewrite the priority arbiter from Example 5.7 using a matching selected assignment as follows:

```
with request select?
  grant <= "1000" when "1---",
          "0100" when "01--",
          "0010" when "001-",
          "0001" when "0001",
          "0000" when others;
```

5.3   If and Case Generate

Earlier versions of VHDL provide two forms of generate statements. A for-generate statement allows us to include multiple copies of component instances or other concurrent statements. An if-generate statement allows us to decide whether or not to include concurrent statements based on the value of a condition. If we want to decide between alternate sets of concurrent statements depending on whether a condition is true or false, we use two if-generate statements with complementary conditions, as follows:

```
L1: if condition generate
    -- first alternative
    ...
end generate L1;

L2: if not condition generate
    -- second alternative
    ...
end generate L2;
```

This is somewhat cumbersome, and inconsistent with sequential if statements, in which we can specify alternates using **elsif** and **else** clauses. VHDL-2008 extends the form of if generate statements to allow us to specify alternatives in a way similar to sequential if statements. We can rewrite the above pair of if-generate statements as follows:

```
L: if condition generate
   -- first alternative
   ...
else generate
   -- second alternative
   ...
end generate L;
```

We can also include further conditions to test, as follows:

```
L: if condition1 generate
   -- first alternative
   ...
elsif condition2 generate
   -- second alternative
```

```
   . . .
   . . .
else generate
  -- last alternative
   . . .
end generate L;
```

Each of the alternatives can be just a set of concurrent statements, or it can include declarations as well as concurrent statements. In the latter case, we write **begin** and **end** keywords around the statements, as follows:

```
L: if condition1 generate
     -- first alternative declarations
     . . .
   begin
     -- first alternative statements
     . . .
   end;
elsif condition2 generate
   . . .
end generate L;
```

When the model is elaborated, the conditions in the if-generate statement are tested from first to last until one is found that is true. The corresponding declarations (if any) and concurrent statements are then included in the elaborated model. If no condition is true and there is an **else** generate alternative, the declarations and statements from that alternative are included. The **else generate** alternative is optional, allowing for the possibility of no declarations or statements being included if none of the conditions is true. Of course, if we omit the **else generate** alternative and there is only one condition to test, the if-generate statement collapses down to the pre-VHDL-2008 form.

EXAMPLE 5.9   *Boundary conditions in a replicated structure*

We often use for-generate statements to replicate cells in a regular structure, and include nested if-generate statements to deal with the differences between the end replications and those in the middle. For example, a ripple-carry adder has a half adder at the least-significant end and has different carry in and out connections for the cells at the ends and in the middle. We can use a nested if-generate with three alternatives to deal with the differences:

```
adder: for i in width-1 downto 0 generate
  signal carry_chain : unsigned(width-1 downto 1);
begin
  adder_cell: if i = width-1 generate -- most-significant cell
    add_bit: component full_adder
      port map (a => a(i), b => b(i), s => s(i),
                c_in => carry_chain(i), c_out => c_out);
```

```
    elsif i = 0 generate -- least-significant cell
      add_bit: component half_adder
        port map (a => a(i), b => b(i), s => s(i),
                  c_out => carry_chain(i+1));
    else generate -- middle cell
      add_bit: component full_adder
        port map (a => a(i), b => b(i), s => s(i),
                  c_in => carry_chain(i),
                  c_out => carry_chain(i+1));
    end generate adder_cell;
  end generate adder;
```

VHDL-2008 also provides a case-generate statement, in which we specify alternatives in a similar way to a case statement. We specify a static expression (one whose value can be computed during elaboration), and choice values for each alternative. The form of a case-generate statement is:

```
L: case expression generate
  when choice1 =>
    -- first alternative
    ...
  when choice2 =>
    -- second alternative
    ...
  ...
end generate L;
```

As in the if-generate statement, each alternative can be just a set of concurrent statements, or it can include declarations as well as concurrent statements, with **begin** and **end** keywords around the statements. The rules governing sequential case statement expressions and choices also apply to the expression and choices in a case-generate statement, with the further stipulation that the expression be static. When the model is elaborated, the expression is evaluated, and the alternative whose choice is the same as the expression value is selected. The declarations (if any) and the statements from that alternative are included in the elaborated model.

EXAMPLE 5.10   *Alternative structures for a complex multiplier*

Multiplication of complex numbers in Cartesian form involves four scalar multiplications, a subtraction, and an addition. Depending on the constraints that apply to a design, these operations can be implemented in one clock cycle using multiple function units, in multiple clock cycles using fewer function units, or in a pipeline. Suppose we have an enumeration type, defined as follows, for specifying the implementation to use:

```
type implementation_type is
        (single_cycle, multicycle, pipelined);
```

An entity declaration for a complex multiplier has a generic constant of this type controlling the implementation:

```
entity complex_multiplier is
  generic ( implementation : implementation_type; ... );
  port ( ... );
end entity complex_multiplier;
```

Within the architecture, we use the value of the generic constant in a case-generate statement to determine what components to instantiate and how to interconnect them:

```
architecture rtl of complex_multiplier is
  ...
begin

  mult_structure : case implementation generate
    when single_cycle =>
        signal real_pp1, real_pp2 : ...;
        ...
      begin
        real_mult1 : component multiplier
          port map ( ... );
        ...
      end;
    when multicycle =>
        signal real_pp1, real_pp2 : ...;
        ...
      begin
        mult : component multiplier
          port map ( ... );
        ...
      end;
    when pipelined =>
        signal real_pp1, real_pp2 : ...;
        ...
      begin
        mult1 : component multiplier
          port map ( ... );
        ...
      end;
  end generate mutl_structure;

end architecture rtl;
```

The case-generate statement includes three alternatives, one for each possible implementation style. Each alternative can have local declarations and concurrent statements with the same names and labels as those in other alternatives, as well as differently named declarations and differently labeled statements.

5.3.1 Configuration of If and Case Generate

One of the main difficulties that has prevented introduction of case-generate statements and if-generate statements with multiple alternatives in earlier versions of VHDL has been working out a way of configuring the alternatives. VHDL-2008 handles this by requiring each alternative to be labeled if it is to be referenced in a configuration declaration. The alternative labels are in addition to the overall statement label. In an if-generate statement, we include a label before each condition:

```
L: if A1: condition1 generate
     -- first alternative declarations
     ...
   begin
     -- first alternative statements
     ...
   end;
elsif A2: condition2 generate
   ...
else A3: generate
   ...
end generate L;
```

We can then use the labels in block configurations for the alternatives within a configuration declaration:

```
for L(A1)
   ...
end for;

for L(A2)
   ...
end for;

for L(A3)
   ...
end for;
```

This is similar to the way in which we write a value or a range in a configuration for a for-generate statement to identify a replication of the generate statement body to configure.

EXAMPLE 5.11   *Configuring a replicated structure with boundary differences*

In Example 5.9 we showed a structure for a ripple carry adder, in which differences among bit positions were handled by alternatives of an if-generate statement. We can revise the statement to include labels in each alternative:

```
adder: for i in width-1 downto 0 generate
  signal carry_chain : unsigned(width-1 downto 1);
begin
  adder_cell: if most_significant: i = width-1 generate
    add_bit: component full_adder
      port map (a => a(i), b => b(i), s => s(i),
                c_in => carry_chain(i), c_out => c_out);
  elsif least_significant: i = 0 generate
    add_bit: component half_adder
      port map (a => a(i), b => b(i), s => s(i),
                c_out => carry_chain(i+1));
  else middle: generate
    add_bit: component full_adder
      port map (a => a(i), b => b(i), s => s(i),
                c_in => carry_chain(i),
                c_out => carry_chain(i+1));
  end generate adder_cell;
end generate adder;
```

We can now write a configuration declaration for the enclosing entity and architecture:

```
configuration widget_cfg of arith_unit is
  for ripple_adder
    for adder

      for adder_cell(most_significant)
        for add_bit: full_adder
          use entity widget_lib.full_adder(asic_cell);
      end for;

      for adder_cell(middle)
        for add_bit: full_adder
          use entity widget_lib.full_adder(asic_cell);
      end for;

      for adder_cell(least_significant)
        for add_bit: half_adder
          use entity widget_lib.half_adder(asic_cell);
      end for;
```

```
    end for; -- adder
  end for; -- ripple_adder
end configuration widget_cfg;
```

The block configuration "**for adder** ... **end for**" configures the for-generate statement. Within it, we have three block configurations, one for each alternative of the if-generate statement. We identify each alternative with a combination of the if-generate statement label (adder_cell) and the alternative label (most_significant, least_significant, and middle, respectively). The configuration information for each alternative is only acted upon during elaboration if the corresponding condition is true and the alternative is included in the design hierarchy.

We handle configuration of alternatives in a case-generate statement in a similar way, by including a label before the choice value or values in each alternative. The form is:

```
L: case expression generate
  when A1: choice1 =>
    -- first alternative
    ...
  when A2: choice2 =>
    -- second alternative
    ...
  ...
end generate L;
```

We also write the configuration information in a similar way, including the label for the alternative in parentheses after the generate statement label.

EXAMPLE 5.12 *Configuring the alternative structures for the complex multiplier*

We can revise the case-generate statement in Example 5.10 to include alternative labels, allowing the alternatives to be configured:

```
mult_structure : case implementation generate
  when single_cycle_mult: single_cycle =>
    ...
  when multicycle_mult: multicycle =>
    ...
  when pipelined_mult: pipelined =>
    ...;
end generate mutl_structure;
```

We can now write a configuration declaration for the complex multiplier:

```
configuration wallace_tree of complex_multiplier is
  for rtl
```

```
      for mult_structure(single_cycle_mult)
        for real_mult1 : multiplier
          use entity work.multiplier(wallace_tree);
        ...
      end for;

      for mult_structure(multicycle_mult)
        for mult : multiplier
          use entity work.multiplier(wallace_tree);
        ...
      end for;

      for mult_structure(pipelined_mult)
        for mult1 : multiplier
          use entity work.multiplier(wallace_tree);
        ...
      end for;

    end for; -- rtl
  end for wallage_tree;
```

The alternative labels in an if-generate or case-generate statement allow us to configure the alternatives of the statement. If we do not need to write an explicit configuration for an alternative, we can leave the alternative unlabeled. In the examples in the first part of Section 5.3, we weren't concerned with configuration for any of the alternatives, so we omitted labels from all alternatives.

Earlier versions of VHDL did not allow for an alternative label in the single alternative of an if-generate statement and did not allow for specification of an alternative label in a corresponding block configuration. VHDL-2008 provides for backward compatibility by allowing a block configuration for an if-generate statement to omit the alternative label and surrounding parentheses. In that case, the block configuration applies to the first alternative of the if-generate statement, and the information in the block configuration is used only if the first condition in the if-generate statement is true.

Chapter 6

Modeling Enhancements

One of the main purposes of VHDL is modeling the behavior of hardware. This chapter describes a number of features in VHDL-2008 that make the modeling task easier. All of the modeling tasks described here can be expressed in earlier versions of VHDL, but not as succinctly.

6.1 Signal Expressions in Port Maps

When we instantiate a component in VHDL, we write a port map to specify the signals connected to the ports of the instance. If an input port is to be tied to a fixed value, we can write an expression in the port map in place of a signal name. In earlier versions of VHDL, the expression was required to be static; that is, the expression's value could not change during execution of the model. A common example is a simple literal value representing tying an unused input high or low:

```
inst : component adder32
  port map ( ..., carry_in => '0', ... );
```

In VHDL-2008, the rules for writing an expression in a port map are generalized to include nonstatic expressions involving the values of signals. This allows us to include a small amount of functional logic in a port map, and avoids the need to express the logic with a separate assignment statement and an intermediate signal. If the expression is not static, the port association is defined to be equivalent to association with an anonymous signal that is the target of a signal assignment with the expression on the right-hand side.

EXAMPLE 6.1   *Select logic in a port map*

Suppose an I/O controller connected to a CPU bus is to be enabled when bus control signals indicate a read from I/O address space and the bus address matches the controller's address. We can include the select logic in the port map for the controller instance:

```
io_ctrl_1 : entity work.io_controller(rtl)
  port map ( en => rd_en and io_sel and addr ?= io_base,
             ... );
```

This is a much more succinct way of expressing the model than the equivalent:

```vhdl
signal en_tmp : std_ulogic;

...

en_tmp <= rd_en and io_sel and addr ?= io_base;

io_ctrl_1 : entity work.io_controller(rtl)
    port map ( en => en_tmp,
                  ... );
```

Most of the time, it is a straightforward matter to determine whether an expression in a port map is static, denoting a fixed value for the port, or nonstatic, implying connection to additional logic. However, if the expression is in the form of a function call applied to a signal name, there are two possible interpretations, one as a nonstatic expression implying connection to logic, and the other as a conversion function implying a change of representation of a value. An example of the latter is:

```vhdl
signal s : signed;
component abstract_ALU is
    port ( a : in integer; ... );
end component;

...

ALU : component abstract_ALU
    port map ( a => to_integer(s), ... );
```

The reason for making the distinction is that the interpretation as a nonstatic expression introduces a delta delay between the input signal changing value and the port changing value, whereas the interpretation as application of a conversion function does not. If we were to write a function with one parameter representing some computational logic, for example:

```vhdl
function increment ( x : unsigned ) return unsigned;
```

and use it in a port map:

```vhdl
op_counter : component reg16
    port map ( d_in => increment(op_count), ... );
```

it would not be clear how to interpret the expression. Under the rules of earlier versions of VHDL, the expression would be interpreted as a conversion function, which is not what we want. To make the intention explicit, we can include the reserved work **inertial** in the port association to imply an inertial signal assignment of the expression to the anonymous intermediate signal. Thus, we would write the port map as

```vhdl
op_counter : component reg16
    port map ( d_in => inertial increment(op_count), ... );
```

Under the VHDL-2008 rules, if we omit the reserved word and the expression can be interpreted as application of a conversion function or a type conversion, then that interpretation takes precedence.

6.2 All Signals in Sensitivity List

The sensitivity list of a process statement specifies the signals that cause the process to resume execution. For a process representing combinational logic, we should include all of the input signals of the logic in the sensitivity list. If we omit an input signal, a synthesis tool would infer latch-based storage, since the output would remain unchanged when the omitted signal changed.

Using earlier versions of VHDL, ensuring that all input signals are included in a sensitivity list is cumbersome and error prone. We must carefully check the body of the process and any subprograms called by the process to assemble the list of signals that are read. If we subsequently revise the model code, we must remember to update each process's sensitivity list if we update the process's statements. In VHDL-2008, writing and maintaining processes representing combinational logic is much simpler. We can replace the list of signals in the sensitivity list with the reserved word **all**, indicating that the process is sensitive to all signals read by the process.

EXAMPLE 6.2 *Combinational logic for a finite-state machine*

One place where assembling the sensitivity list for a combinational process often causes problems is the next-state and output logic for a finite-state machine. The logic has, as inputs, the current state signal and signals whose values determine the next state and the output values. An example is:

```
next_state_logic : process (all) is
begin
  out1 <= '0'; out2 <= '0'; ...
  case current_state is
    when idle =>
      out1 <= '1';
      if in1 and not in2 then
        out2 <= '1';
        next_state <= busy1;
      elsif in1 and in2 then
        next_state <= busy2;
      else
        next_state <= idle;
      end if;
    ...
  end case;
end process next_state;
```

As we revise the finite-state machine, we might include more signals as inputs. Using the reserved word **all** instead of explicitly listing the input signals makes the process easier to write and maintain.

The list of signals to which the process is made sensitive also includes signals read in subprograms called directly or indirectly by the process. A proviso, however, is that a subprogram declared in a separate design unit from the process must only read its signal parameters. It cannot read signals declared globally in packages or signals identified using external names (see Section 2.1). The rationale for this restriction is to ensure the analyzer can assemble the sensitivity list for the process just by analyzing the enclosing design unit. It also means that we can perform a similar analysis to understand a process. We don't get any surprises from unexpected external signals becoming inputs to the logic implied by a process.

6.3   Reading Out-Mode Ports and Parameters

In many designs, we compute a value and assign it to an output port, and then use the value in further computations within the design. There are two ways in which we might do this. On one hand, the value might be used to implement further behavior of the design. An example is a circuit with both active-high and active-low outputs. We derive a value for the active-high output, and then negate it for the active-low output. One the other hand, the value might be used in verification of the design's functionality. The computation using the value is passive and does not imply additional hardware.

VHDL provides two modes for output ports, **out** and **buffer**. An **out**-mode port is intended to be used for cases where there is no internal use of the port's value. Prior to VHDL-2008, we could not read the value of an **out**-mode port. A **buffer**-mode port is intended to be used for cases where the port's driving value is used internally. As the name suggests, a hardware buffer is implied between the driver assigning values to the port and the external connections. The value used internally is taken from the inside of the buffer.

Prior to VHDL-2002, there were somewhat restrictive rules governing connection of **buffer**-mode ports of a component instance to the **buffer**-mode and **out**-mode ports of an enclosing entity. These restrictions made it very difficult to use **buffer**-mode ports effectively, so designers largely ignored them. It became common practice to declare an internal signal in a design, to use that signal internally, and to include a separate signal assignment to drive the signal's value onto an **out**-mode port.

In VHDL-2002, the rules for **buffer**-mode ports were relaxed, allowing them to be connected to external **out**-mode and **buffer**-mode ports. It is no longer necessary to resort to an internal signal when a port's driving value is needed internally as an input to further logic. Nonetheless, the practice persists, both for backward compatibility with previous versions of VHDL, and because many designers are not aware of the change.

EXAMPLE 6.3   *True and complement outputs for a flip-flop*

A flip-flop having both true (active-high) and complement (active-low) outputs can be modeled using **out**-mode ports and an internal signal as follows:

```
entity Dff is
  port ( clk, d : in bit; q, q_n : out bit );
end entity Dff;

architecture rtl of Dff is
  signal q_int : bit
begin
  ff : process (clk) is
  begin
    if clk = '1' then
      q_int <= d;
    end if;
  end process ff;
  q   <=     q_int;
  q_n <= not q_int;
end architecture rtl;
```

By deriving the active-low output from the active-high internal signal, we avoid having a synthesis tool infer two separate flip-flops. Under the relaxed rules for **buffer**-mode ports in VHDL-2002, we can rewrite this model as follows:

```
entity Dff is
  port ( clk, d : in bit; q : buffer bit; q_n : out bit );
end entity Dff;

architecture rtl of Dff is
begin
  ff : process (clk) is
  begin
    if clk = '1' then
      q <= d;
    end if;
  end process ff;
  q_n <= not q;
end architecture rtl;
```

In VHDL-2008, the restriction on reading **out**-mode ports is also removed. The value of an **out**-mode port seen internally is the same as the value being driven onto the port. Thus, an **out**-mode port has the same behavior in VHDL-2008 as a **buffer**-mode port. However, the rationale for allowing reading of an **out**-mode port is to support verification of a design's behavior, as opposed to implying a hardware buffer. We should choose between **out** and **buffer** modes for a port depending on whether we intend to imply hardware buffering or passive verification, respectively. The choice of port mode documents our intention. Note that this choice would be made as a convention, rather than being enforced by the language definition. A tool has no way of distinguishing between the two ways of using a port.

EXAMPLE 6.4   *Reading an out-mode port for verification*

Suppose we wish to verify that the outputs of a device are all 'Z' within a required interval of the device being disabled and remain all 'Z' until the device is enabled. The output values are not required internally to implement any functionality for the device. Hence, we declare the output ports using **out** mode, as follows:

```
entity device is
  generic ( T_z : delay_length );
  port ( en : in std_ulogic;
          d_out : out std_ulogic_vector(7 downto 0); ... );
end entity device;
```

We can read the values driven onto the output ports in verification code in the architecture:

```
architecture verifying of device is
begin
  d_out <= ... when to_x01(en) = '1' else
            ... when to_x01(en) = '0' else
           "XXXXXXXX";
  assert (to_x01(en'delayed(T_z)) = '0' and d_out = "ZZZZZZZZ")
          or to_x01(en) = '1';
end architecture verifying;
```

VHDL-2008 also removes the restriction on reading the value of an **out**-mode parameter of a procedure. **Out**-mode parameters can be either signal parameters or variable parameters. In the case of an **out**-mode signal parameter, when we call the procedure and pass an actual signal, the procedure is passed a reference to the actual signal. The value we get when we read the parameter within the procedure is the current value of the actual signal.

In the case of an **out**-mode variable parameter, the rules are slightly different. The formal parameter of the procedure is treated like a local variable, and is initialized when the procedure is called. It does not take on the value of the actual parameter variable at that stage. As we make assignments to the parameter within the procedure, the local value is updated. When we read the parameter, we get that local value. Eventually, when the procedure returns, the final value of the parameter is copied to the actual variable.

To illustrate this behavior, consider the following procedure:

```
procedure do_funny ( p : out positive ) is
  variable tmp : positive;
```

```
begin
  tmp := p;
  p := tmp * 2;
end procedure do_funny;
```

Suppose we call the procedure as follows:

```
variable v : positive := 10;
...

do_funny(v);
```

When the procedure is called, the parameter **p** is initialized, not with the value of **v**, but with the default initial value for **positive**, namely, 1. Thus, the value read and assigned to **tmp** is 1, and the value assigned to **p** in the last statement is 2. When the procedure returns, the final value of **p**, namely, 2, is copied out to **v**.

EXAMPLE 6.5   *Checking an out-mode status parameter*

A procedure to read multiple items from a line of input might have an **out**-mode parameter to indicate whether items were read successfully. We can use the parameter as the actual to **textio read** procedures, and check that each call succeeded before continuing, as follows:

```
procedure read_time_and_int ( L : inout line;
                              T : out time; I : out integer;
                              good : out boolean ) is
begin
  read(L, T, good);
  if not good then
    report "Read of time failed" severity failure;
    return;
  end if;
  read(L, I, good);
  if not good then
    report "Read of integer failed" severity failure;
  end if;
end procedure read_time_and_int;
```

If the procedure could not read the value of the **out**-mode parameter **good**, it would have to declare an internal variable to pass to the **read** procedures, and copy the value to **good** upon returning. The ability to read the **out**-mode parameter makes the procedure simpler and easier to maintain.

6.4   Slices in Aggregates

An array aggregate provides a way of forming an array from a collection of elements. In earlier versions of VHDL, we could only form an aggregate from individual elements. In VHDL-2008, the rules are extended to allow us to form an aggregate from a mixture of individual elements and slices of the array. For example, a **bit_vector** array aggregate could be written as follows:

```
('0', "1001", '1')
```

This forms a 6-element vector from the single element '0', the vector value "1001" and the single element '1'. The vector value forms a slice of the final aggregate value. The effect is similar to concatenating the element and array values. While this illustrates the idea, a more powerful use of the feature involves writing an aggregate as the target of an assignment statement. We can write a signal assignment target in the form of an aggregate of signal names. In VHDL-2008, the names can be a mixture of element-typed signals and array-typed signals. Elements of the right-hand-side value are assigned to the matching signals or signal elements.

EXAMPLE 6.6   *Binary addition with carry out*

We can add two 16-bit unsigned numbers to get a 17-bit result, with the most-significant bit being the carry out of the addition, and the least-significant 16 bits being the sum. We can assign to separate signals representing the carry out and the sum as follows:

```
signal a, b, sum : unsigned(15 downto 0);
signal c_out     : std_ulogic;
...

(c_out, sum) <= ('0' & a) + ('0' & b);
```

The operands **a** and **b** are zero-extended and added to produce a 17-bit result. This is assigned to the aggregate target comprising the scalar signal **c_out** and the 16-bit vector **sum**. Since we use positional association in the aggregate, **c_out** matches the leftmost element and **sum** matches the rightmost 16 elements.

We can also use named association in array aggregates. If we want to include an association for a slice of the aggregate, we specify a range of index values for the slice. For example, given signals declared as follows:

```
variable status_reg : bit_vector(7 downto 0);
variable int_priority, cpu_priority : bit_vector(2 downto 0);
variable int_enable, cpu_mode : bit;
```

We can write the assignment:

```
( 2 downto 0 => int_priority,
  6 downto 4 => cpu_priority,
  3 => int_en, 7 => cpu_mode )    := status_reg;
```

This specifies that the bits of the **status_reg** value are assigned in left-to-right order to **cpu_mode**, **cpu_priority**, **int_en**, and **int_priority**, respectively. When we include slices in an aggregate, the direction of the aggregate's index range is taken from the direction of ranges used to specify the slices. Thus, in this above example, the direction is descending, since "2 **downto** 0" and "6 **downto** 4" are both descending ranges. All the ranges used in this way must have the same direction. Thus, it would be illegal to write the example using "2 **downto** 0" for one range and "4 **to** 6" for the other.

6.5   Bit-String Literals

VHDL allows us to use a bit-string literal to specify a value for a vector of '0' and '1' elements in binary, octal, or hexadecimal form. The rules for bit-string literals in earlier versions of VHDL were somewhat restrictive. In particular, the resulting vector value had to be a multiple of three in length for octal literals or a multiple of four in length for hexadecimal literals. Moreover, every bit had to be stated explicitly, as there was no provision for zero extension or sign extension. Finally, only '0' and '1' elements could be specified. There was no provision for "metalogical" values, such as 'X' or 'Z'.

VHDL-2008 enhances bit-string literals considerably. First, we can specify elements other than just '0' and '1'. In an octal literal, any non-octal-digit character is expanded to three occurrences of that character in the vector value. Thus, the literal O"3XZ4" represents the vector value "011XXXZZZ100". Similarly, in a hexadecimal literal any non-hexadecimal-digit character is expanded to four occurrences of the character. Thus, the literal X"A3—" represents the vector value "10100011————". In a binary literal, any non-bit character just represents itself in the vector. Thus, B"00UU" represents the vector "00UU". While this may seem vacuous at first, the benefit of allowing this in binary literals will become clear when we look at other enhancements.

Note that expansion of non-digit characters does not extend to embedded underscores, which we might add for readability. Thus, O"3_X" represents "011XXX", not "011___XXX". Also, expansion on non-digit characters is not limited to those defined in std_ulogic, though that is the most common use case. We could write the literal X"0#?F" to represent the string value "0000####????1111".

The second enhancement is provision for specifying the exact length of the vector represented by a literal. This allows us to specify vectors whose length is not a multiple or three (for octal) or four (for hexadecimal). We do so by writing the length immediately before the base specifier, with no intervening space. For example, the literal 7X"3C" represents the 7-element vector "0111100", 8O"5" represents "00000101", and 10B"X" represents "000000000X". If the final length of the vector is longer than that implied by the digits, the vector is padded on the left with '0' elements. If the final length is less than that implied by the digits, the left-most elements of the vector are truncated, provided they are all '0'. An error occurs if any non-'0' elements are truncated, as they would be in the literal 8X"90F".

The third enhancement is provision for specifying whether the literal represents an unsigned or signed number. We represent an signed number using one of the base specifiers UB, UO, or UX. These are the same as the ordinary base specifiers B, O, and X. When a sized unsigned literal is extended, it is padded with '0' elements, and when elements are truncated, they must be all '0'.

We represent a signed number using one of the base specifiers SB, SO, or SX. The rules for extension and truncation are based on those for sign extension and truncation of 2s-complement binary numbers. When a sized signed literal is extended, each element of padding on the left is a replication of the leftmost element prior to padding. For example, the literal 10SX"71" is extended to "0001110001", 10SX"88" is extended to "1110001000", and 10SX"W0" is extended to "WWWWWW0000". When a sized signed literal is truncated, all of the elements removed at the left must be the same as the leftmost remaining element. For example, the literal 6SX"16" is truncated to "010110", 6SX"E8" is truncated to "101000", and 6SX"H3" is truncated to "HH0011". However, 6SX"28" is invalid, since, prior to truncation, the vector would be "00101000". The two leftmost elements removed are each '0', which differs from the leftmost remaining '1' element. The literal would have to be written as 6SX"E8" for this reason. The rationale for this rule is that it prevents the signed numeric value represented by the literal being inadvertently changed by the truncation.

The remaining enhancement is provision for specifying a vector value in decimal. For this, we use the base specifier D. All of the characters in the literal must then be decimal digits (or underscores); we cannot specify other characters, such as 'Z' or 'X', since it would not be clear which elements of the vector would correspond to those characters. If we omit a size specification in a decimal bit-string literal, the number of elements represented is the smallest number that can encode the value. For example, the literal D"23" represents the vector "10111", D"64" represents "1000000", and D"0003" represents "11". A decimal bit-string literal is treated as representing an unsigned number. If the literal must be extended, the vector is padded on the left with '0' elements. For example, 12D"10" represents "000000001010". It can never be legal to specify a size requiring truncation, since the leftmost element prior to truncation is always '1'.

Chapter 7

Improved I/O

Previous versions of VHDL provided fairly basic forms of binary and text I/O. In this chapter, we describe additions to the I/O and string conversion features provided by VHDL-2008.

7.1 The **To_string** Functions

In previous versions of VHDL, converting a value to a string required the use of the 'image attribute or the **write** procedures from the **textio** package. These options were very limiting. The 'image attribute was only defined for scalar types, and the **write** procedures were only defined for the predefined scalar types and for **bit_vector**. VHDL-2008 adds predefined **to_string** operations as a flexible alternative for string conversion. As a function, **to_string** is overloadable, supports scalar and composite types, and can have multiple parameters. In addition, for all bit-based array types, octal and hexadecimal string conversion functions are defined: **to_ostring** and **to_hstring**, respectively (see Section 7.1.3).

> **EXAMPLE 7.1** *Writing messages using to_string*
>
> ---
>
> One use of these functions is in assert statements, for example:
>
> ```
> assert expected_val = read_val
>   report "Expected Val /= Actual Val." &
>          "   Expected = " & to_string(expected_val) &
>          "   Actual = "   & to_string(read_val)
>   severity error;
> ```
>
> Another use is with VHDL's built-in **write** procedure (not **std.textio.write**) as follows:
>
> ```
> if expected_val = read_val then
>   err_cnt := err_cnt + 1;
>   write(output,
>     "%%%Error:  Expected Val /= Actual Val." &
>     "   Expected = " & to_string(expected_val) &
>     "   Actual = "   & to_string(read_val) &
> ```

```
    "   at time: "   & to_string(now));
end if;
```

This call to the **write** procedure has a similar effect to a sequence of calls to the **write** procedures defined in the **textio** package, followed by a call to the **writeline** procedure. However, it is clearly much more concise.

7.1.1 Predefined **To_string** Functions

A basic form of **to_string** is predefined with the following signature.

```
to_string[AType return string]
```

AType includes all scalar types and single dimensional array types whose element types contain only character literals. The string return value for various types is formed as follows:

- For a value of an enumeration type other than **character**, if the value is a character literal, **to_string** returns the value as a single-element string; otherwise, the function returns the name of the identifier in lower case letters in a string. For example,

    ```
    to_string(bit'('0'))
    ```

    returns the string "0", and

    ```
    to_string(file_open_status'(OPEN_OK))
    ```

    returns the string "open_ok".

- For a **character** value, **to_string** returns the character in a single-element string. Note that this may be different from the result for non-character types. In the case of control-character values, the result is a single-element string containing the control character, not a string containing the name of the control character.

- For a one-dimensional array value containing only character literals, **to_string** returns a string of the same length as the array containing the element values converted to type **character**. For example, **to_string**(bit_vector'("0110")) returns the string of characters "0110".

- For a value of an integer type, **to_string** returns the decimal literal. There is no exponent and no insignificant leading zeros. If the result is negative, the decimal literal is preceded immediately by a minus sign without any intervening space.

- For a value of a floating point type, including **real**, **to_string** returns the decimal literal in standard form consisting of a normalized fraction and an exponent in which the sign is present and the "e" is in lowercase. There are no insignificant leading or trailing zeros. (Note that the floating point types referred to here are the abstract numeric types, not the types defined in the new floating-point packages described in Section 8.5.)

- For a value of a physical type, **to_string** returns the decimal literal as an integer, a space, and the unit name. There is no exponent and no insignificant leading zeros. If the result is negative, the decimal literal is preceded immediately by a minus sign without any intervening space.

7.1.2   Overloaded **To_string** Functions

The basic **to_string** operations handle most of the required cases. However, there are additional predefined forms of **to_string** for values of types **time** and **real**. First, the following predefined forms provide **textio**-style formatting:

```
function to_string ( value: time; unit : time )
                    return string;

function to_string ( value: real; digits: natural )
                    return string;
```

The version for type **time** formats the value as an integer or real literal in multiples of **unit**. In the version for type **real**, **digits** specifies the number of digits that are to appear to the right of the decimal point.

Second, for type **real**, the following predefined form of **to_string** provides C-style formatting of the value:

```
function to_string ( value: real;  format: string )
                    return string;
```

The **format** parameter should contain a C-style format specification, such as is used in the C **printf** command. For example, assuming the variable **x** has the value 52.5:

- to_string (value => x, format => "%f") returns "52.5"

- to_string (value => x, format => "%5.2f") returns "52.50"

- to_string (value => x, format => "%E") returns "5.250000E+01"

- to_string (value => x, format => "%6.2e") returns "5.25e+01"

- to_string (value => (x*10.0), format => "%g") returns "525"

The basic **to_string** operations are predefined for the vector types defined in packages std_logic_1164, numeric_bit and **numeric_std**, since the types just contain character elements ('0', '1', 'L', 'H', and so on). However, the fixed-point and floating-point packages described in Chapter 8 overload the basic operations to provide more useful results. In particular, the overloaded versions defined for the fixed-point types **ufixed** and **sfixed** return a string with a radix point (a '.' character) at the appropriate position, for example, "1001.0010". The overloaded version for the floating-point type **float** returns a string with colon characters between the sign, exponent, and fraction bits, for example, "0:111011:00010001110".

7.1.3   The To_ostring and To_hstring Functions

In addition to binary string conversion, VHDL-2008 also adds octal and hexadecimal string conversion functions, **to_ostring** and **to_hstring**, respectively, for all bit-based array types. The signatures for these functions are:

```
to_ostring[BitArrayType return string]

to_hstring[BitArrayType return string]
```

BitArrayType includes **bit_vector** defined in the package **standard**, **std_ulogic_vector** defined in **std_logic_1164**, **unsigned** and **signed** defined in **numeric_std**, and **unsigned** and **signed** defined in **numeric_bit**. For these types, characters are implicitly added to the left of the array value to make the length a multiple of 3 (for **to_ostring**) or 4 (for **to_hstring**), so that complete octal or hexadecimal digits can be formed. The characters are added as follows:

- For an array of type **bit_vector**, '0' characters are added.

- For an array of type **std_ulogic_vector**, **std_logic_vector**, or **unsigned**, if the leftmost element of the array is 'Z' or 'X', then 'Z' or 'X' characters, respectively, are added; otherwise, '0' characters are added.

- For an array of type **signed**, the characters added are the same as the leftmost element of the array.

For array types based on either **std_ulogic** or **std_logic**, if all of the elements corresponding to an octal or hexadecimal digit contain 'Z' then the resulting character is 'Z'. Otherwise, **to_X01** is applied to the group of elements. If the result contains an 'X', the octal or hexadecimal digit is 'X'. If the result contains only '0' and '1' values, they are converted to a normal octal or hexadecimal digit in upper case.

BitArrayType is supported for the fixed-point types **ufixed** and **sfixed** defined in the fixed-point package (see Section 8.4). For these types, characters before the radix point are handled as for **unsigned** and **signed**, respectively. Then the radix point is included in the string. For characters following the radix point, '0' characters are added to the right of the array value to make the length of the fractional part (after the radix point) a multiple of 3 (for **to_ostring**) or 4 (for **to_hstring**). The fractional part is then converted as for **unsigned** values. For values in which the radix-point position lies outside the index range, **to_ostring** and **to_hstring** extend the value to include the radix point in the result. For example, a **to_hstring** operation for the value "10100" with index range 7 down to 3 would result in the string "A0.0", corresponding to the binary number 10100000.0. Similarly, a **to_hstring** operation for the value "10100" with index range –3 down to –7 would result in "0.28" (0.0010100 in binary).

BitArrayType is also supported for type **float** in the floating-point package (see Section 8.5). For this type, characters are added to the left of the array value string to make the length a multiple of 3 (for **to_ostring**) or 4 (for **to_hstring**). If the left most element of the array is 'Z' then 'Z' characters are added; otherwise, '0' characters are added.

Note that for consistency, there is also a **to_bstring** operation defined for each of the types. However, it is simply an alias for the **to_string** function. There are further aliases defined for the operations: **to_binary_string**, **to_octal_string**, and **to_hex_string**. Some designers may consider these to be more readable than the shorter function names.

7.2 The Justify Function

One facility provided by the write operations in the **textio** package but not provided by the **to_string** functions is the ability to justify a string representation in a field of a given width. This is useful for tabular formatting of output. VHDL-2008 adds a **justify** function to the **textio** package to provide fixed-width formatting for string values. The function is defined as follows:

```
function justify ( value    : string;
                   justified : side  := right;
                   field    : width := 0 ) return string;
```

The **value** parameter contains the string value to be formatted, and the **justified** and **field** parameters are used in the same way as in the **write** procedures in **textio**. When the **field** parameter is greater than the length of **value**, the value is justified within the string by adding spaces to the left or right of the result for right, depending on the **justified** parameter. If the **field** parameter is less than or equal to the length of **value**, the value is returned unchanged.

EXAMPLE 7.2 *Tabular formatting of trace output*

Suppose we wish to write trace output from a model and have it formatted in fixed-width columns. Each line of output consists of the current simulation time, a 16-bit **unsigned** value in hexadecimal format, and an integer counter value. We can write the values using the following **write** procedure call:

```
write( output, justify(to_string(now, ns),  width => 10),
                justify(to_hstring(out_vec), width =>  6),
                justify(to_string(count),    width => 10) );
```

Successive calls might yield the following output:

```
 20 ns  XXXX        0
120 ns  ZZ00        1
220 ns  FFC0       10
320 ns  0000       31
```

7.3 Newline Formatting

In some applications, we need to create a message string that would be too long to fit on a single line of output. If we are using **write** or **writeline** operations, we could split the message and write each line with a separate operation. However, VHDL does not guarantee that lines written by different processes during the same simulation cycle will not be interleaved. If we are generating a message using an assert or report statement, we would not want the additional text associated with an assertion violation to be included with each line of the message.

VHDL-2008 allows us to create multiline messages in these cases by using the line-feed character (LF) as a newline character. This interpretation applies to a report string in an assert or report statement and to a string written to a file of type **text** using a **write**, **writeline**, or **tee** procedure (see Section 7.5). The host operating system translates the LF character to whatever convention is used to represent a new line. For example, a UNIX based system would represent the new line using just the LF character, whereas a Windows system would represent it using a carriage return (CR) followed by a LF.

EXAMPLE 7.3 *Multiline output to a text file*

We can use LF characters in a multiline message written to the standard **output** text file using a **write** procedure call:

```
write(output, "%%%ERROR data value miscompare." & LF &
              " Actual value = "   & to_hstring(data) & LF &
              " Expected value = " & to_hstring(expdata) & LF &
              " at time:   "       & to_string(now) );
```

7.4 Read and Write Operations

In earlier versions of VHDL, textual I/O operations were limited to values of predefined types, for which **read** and **write** operations were defined in the standard **textio** package. VHDL-2008 broadens the support for textual I/O by adding operations for all of the standard types in their respective packages. It also adds octal and hexadecimal I/O and enhances the string I/O capability.

The basic I/O operations added in each package are:

```
procedure write ( L : inout line;  value : in AType;
                  justified : in side := right;
                  field : in width := 0 );

procedure read  ( L : inout line;  value : out AType;
                  good : out boolean );

procedure read  ( L : inout line;  value : out AType );
```

These operations behave in the same way as the corresponding operations for predefined types in the **textio** package. The **write** operation executes as if the following call to the **textio** package **write** operation were executed with the **to_string** operation (see Section 7.1) applied to the parameter:

```
write (L, to_string(value), justified, field);
```

The new write operations are defined for std_ulogic_vector and std_logic_vector in the std_logic_1164 package, for unsigned and signed in numeric_std, for unsigned and signed in numeric_bit, for ufixed and sfixed in the fixed-point packages (see Section 8.4), and for float in the floating-point packages (see Section 8.5).

Each read procedure skips white space, and then reads std_logic values until it encounters white space or a non-std_ulogic value, or until it has read value'length characters. Underscore characters ("_") embedded within the value are skipped, though it is an error if two underscores appear consecutively. The procedure must read enough characters to fill all of the elements of the value array, so it is an error if a space or an invalid character is encountered before value'length characters are read. The read procedures for ufixed and sfixed also accept a radix point (".") in the input, though it is an error if the radix point is not at the appropriate position. Specifically, the characters before the radix point must fill elements of the value parameter with non-negative indices, and the characters after the radix point must fill elements with negative indices. An error occurs if the radix point is encountered at a position other than between the characters corresponding to indices 0 and –1. Similarly, the read procedures for float accept ":" and "." delimiters between the sign, exponent, and fraction parts of the input, though it is an error if they are not at the appropriate positions.

The support for octal and hexadecimal I/O takes the form of the following procedures:

```
procedure owrite ( L : inout line;  value : in AType;
                   justified : in side := right;
                   field : in width := 0 );

procedure hwrite ( L : inout line;  value : in AType ;
                   justified : in side := right;
                   field : in width := 0 );

procedure oread ( L : inout line;  value : out AType;
                  good : out boolean );

procedure oread ( L : inout line;  value : out AType );

procedure hread ( L : inout line;  value : out AType;
                  good : out boolean );

procedure hread ( L: inout line;  value : out AType );
```

These operations also behave in the same way as the corresponding operations for predefined types in the textio package, but with octal or hexadecimal conversion applied. The owrite and hwrite operations execute as if the following calls to the textio package write operation were executed with the to_ostring and to_hstring operations (see Section 7.1) applied to the parameters:

```
write (L, to_ostring (value), justified, field);  -- owrite

write (L, to_hstring (value), justified, field);  -- hwrite
```

The operations are defined for the predefined type **bit_vector**, for **std_ulogic_vector** and **std_logic_vector** defined in the **std_logic_1164** package, for **unsigned** and **signed** in **numeric_std**, and for **unsigned** and **signed** in **numeric_bit**. The behavior of the **oread** and **hread** operations in these cases is as follows. Each operation must read sufficient characters to fill the **value** argument, or an error occurs. Since **value** need not be a multiple of 3 (for **oread**) or 4 (for **hread**) in length, the length is rounded up to the nearest multiple of 3 or 4 to determine how many characters to read. **Oread** (**hread**) starts by skipping white space. It then reads octal (hexadecimal) digits until it encounters white space or a non-octal (non-hexadecimal) character other than "_", or until it has read sufficient characters to fill the **value** argument. Underscore characters embedded within the octal (hexadecimal) value are skipped. **Oread** converts each octal digit (0–7) to its 3-bit representation, and **hread** converts each hexadecimal digit (0–9, a–f, or A–F) to its 4-bit representation. For array types based on **std_ulogic** or **std_logic**, the characters 'X' and 'Z' are also permitted. For octal, these characters are repeated 3 times in the result; hence, a 'Z' input is expanded to "ZZZ". Similarly, for hexadecimal, these characters are repeated 4 times in the result; hence, a 'Z' input is expanded to "ZZZZ". If conversion of characters to groups of 3 or 4 elements result in more elements than the length of the **value** argument, only the rightmost elements are used. Depending on the values of the discarded elements, an error may occur. If the type of the **value** argument is **bit_vector**, **std_ulogic_vector**, **std_logic_vector**, or **unsigned**, an error occurs if any of the discarded elements are '1'. For example, an **hread** that reads the characters "82" ("10000010" in binary) into a 6-bit **unsigned** value produces an error, since the two discarded bits are "10". If the type of the value argument is **signed**, an error occurs if the discarded elements are not all the same as the leftmost element used for the **value** argument. For example, an **hread** that read the characters "7F" into a 6-bit **signed** value produces an error, since the two discarded bits are "01", and the leftmost bit used for **value** is '1'.

The octal and hexadecimal write and read operations are also defined for types **ufixed** and **sfixed** in the fixed-point packages (see Section 8.4). The behavior of the **oread** and **hread** operations in these packages is as follows. **Oread** and **hread** each reads the value prior to the radix point as described above for **unsigned** or **signed** (depending on whether the **value** parameter is **ufixed** or **sfixed**, respectively). For the characters following the radix point, **oread** and **hread** each reads the value as described above for **std_ulogic_vector**; however, instead of discarding elements on the left, the operations discard elements on the right. An error occurs if an element discarded on the right is a '1'. The radix point may be explicitly included in the input, but an error occurs if it is not at the appropriate position (that is, between the characters corresponding to indices 0 and − 1 of the **value** parameter). The radix point may also be omitted, in which case it is assumed at the appropriate position.

Finally, the octal and hexadecimal write and read operations are defined for the type **float** in the floating-point packages (see Section 8.5). The behavior of the **oread** and **hread** operations depends on whether ":" or "." delimiters are used in the input to separate the sign, exponent, and fraction parts of a floating-point number. When ":" delimiters are used (with the input formatted as "S:EEEE:FFFFFFFF"), the sign bit, the exponent, and the fraction are each read as separate octal or hexadecimal values using the same rules as described above for **std_ulogic_vector** values. When a '.' delimiter is used (with the input formatted as "SEEEE.FFFFFFFF"), the rules described above for reading **ufixed** values are used. The value read before the radix point forms the part of the result comprising the sign and exponent elements, and the value read after the radix point forms the fraction part of the result. When no delimiters are used in the input, the entire **float** value is read as a single hexadecimal value as described above for **std_ulogic_vector** values.

Note that, for consistency, there are also definitions of binary I/O operations in each of the packages. However, they are simply aliases for the basic operations, defined as follows:

```
alias bwrite is write [line, AType, side, width];

alias bread is read [line, AType, boolean];

alias bread is read [line, AType];
```

In addition to the enhanced I/O operations for binary, octal, and hexadecimal values, VHDL-2008 adds enhanced I/O operations for character strings. The new **swrite** procedure in package **textio** writes a string value in the same way as the **write** procedure overloaded for a **string** value parameter. The difference is that there are no other overloaded versions of **swrite**, so we do not have to use type qualification when writing a **string** literal. We can write a call such as:

```
swrite(L, "The answer is: ");
```

Compare this with a call to the **write** operation:

```
write(L, string'("The answer is: "));
```

We need to use the type qualification to distinguish the type of the string from other character array types (such as **bit_vector**) for which **write** is defined. Use of **swrite** to write **string** literals makes the model much clearer.

The new **sread** procedure in package **textio** reads string-based tokens. It is defined as follows:

```
procedure sread ( L : inout line;
                  value : out string; strlen : out natural );
```

The procedure skips leading white space and then reads consecutive non-white space characters, up to as many as will fit in the **value** parameter. The number of characters read is returned in the **strlen** parameter. If a white-space character stops reading before **value** is filled, the result in **strlen** will be less than the length of **value**. The remaining unfilled characters in **value** are not specified and should not be used. If no valid characters are read (for example, if the input is blank up to the end of the line), the **strlen** result value is 0.

7.5   The Tee Procedure

The **textio** package in previous versions of VHDL provided the file **output** for displaying messages to a user-interface. If we also wanted to log the messages in a separate file for subsequent analysis, we had to duplicate the write operations: once to **output** and once to the separate file. VHDL-2008 adds a **tee** procedure to the **textio** package that writes a line both to the file **output** and to a separate named file. This allows us to avoid replicated write operations. The definition of **tee** is:

```
procedure tee (file F : text; L : inout line);
```

The effect of TEE is the same as the statements:

```
write     (output, L.all & LF);
writeline (F, L) ;
```

EXAMPLE 7.4   *Logging output messages*

Suppose we wish to write trace messages to a simulator's user interface and to log the messages to a file named **trace.log**. We can do this using calls to **tee** in place of writeline, as follows:

```
use std.textio.all;
file tracefile : text open write_mode is "trace.log";
variable L : line;
...

swrite(L, justify(to_string(now, ns), field => 10) &
          " starting operation ");
tee(tracefile, L);
...
```

7.6   The **Flush** Procedure

VHDL-2008 adds a predefined file **flush** procedure that requests that the effect of all previous calls to the **write** procedure for a file be completed. The procedure **flush** is predefined for all file types as follows:

```
procedure flush ( file F : FT );
```

When the **flush** procedure is called, the file must be opened in write or append mode; otherwise, an error occurs.

One use of **flush** is to ensure that all outstanding write operations to an external file are completed before read operations are performed on a separate file object associated with the same external file. Another use is to ensure that prompt messages written to a user interface appear before read operations take input from the user interface. It is important to note, however, that the **flush** operation simply requests that the host operating system complete the outstanding writes. It does not guarantee that the request will be met. In particular, host systems that use distributed network file systems may find it difficult to reliably honor flush requests.

Chapter 8

Standard Packages

In earlier versions of VHDL, the predefined types were declared in the package **standard**, specified in the VHDL Standard Language Reference Manual (LRM). Other standard types were specified in packages defined by separate IEEE standards. They included the packages **math_real**, **math_complex**, **std_logic_1164**, **numeric_bit**, and **numeric_std**. In VHDL-2008, all of these packages are included as part of the VHDL LRM, and so are now considered to be part of VHDL. VHDL also adds a number of new packages, including packages for fixed-point and floating-point numbers represented as vectors of **std_ulogic** elements, and a package providing access to the simulation environment. Furthermore, VHDL-2008 adds operations to the standard packages to provide a consistent feature set across the suite. This includes a consistent set of conversion functions and the I/O operations **to_string**, **read**, **write**, and so on. In this chapter, we summarize the contents of the packages.

8.1 The **Std_logic_1164** Package

The **std_logic_1164** package defines the types **std_ulogic**, **std_logic**, **std_ulogic_vector** and **std_logic_vector**, as well as operations on these types. VHDL-2008 makes the following enhancements to the package:

- In earlier versions, **std_ulogic_vector** and **std_logic_vector** were declared as separate types. That meant many of the operations had to be declared in two overloaded forms, one for each type. It also made it difficult for us to mix the two types in designs where some signals had multiple resolved sources and others had only a single source. In VHDL-2008, the **std_logic_1164** package is revised to take advantage of the new features for resolving elements of composite types (see Section 3.2). The type **std_logic_vector** is now a subtype of **std_ulogic_vector**. Each of the array operations in the package is defined just for the **std_ulogic_vector** type and can be applied to **std_logic_vector** values.

- The VHDL-2008 version of the package defines array/scalar logic operations for **std_ulogic** and **std_ulogic_vector** values, mirroring those that are predefined for **bit/bit_vector** and **boolean/boolean_vector** (see Section 4.1).

- The package adds logical reduction operations for **std_ulogic_vector** values (see Section 4.3).

- The matching relational operators (see Section 4.5) are predefined: "?=", "?/=", "?>", "?>=", "?<", and "?<=" for **std_ulogic**, and "?= "and "?/=" for **std_ulogic_vector**.

- The **maximum** and **minimum** functions are predefined for **std_ulogic** and **std_ulogic_vector** (see Section 4.6).

- The package defines overloaded shift operations (**sll**, **srl**, **rol**, **ror**) for **std_ulogic_vector**. It does not add **sra** or **sla**, since those operations assume a numeric interpretation for a vector. Overloaded version of the arithmetic shift operations are added to **numeric_std_unsigned** instead (see Section 8.3).

- The condition operator ("??") is defined for **std_ulogic**, allowing logical expressions that yield **std_ulogic** values to be used as Boolean conditions (see Section 4.4).

- The package defines a complete set of string conversion functions and text I/O procedures (see Chapter 7). Many of these operations provide the same functionality as operations in the non-standard **std_logic_textio** package provided by some tool vendors. To ease the transition from that package to the standard packages, VHDL-2008 provides an empty version of **std_logic_textio** package. Legacy code that included a use clause referring to **std_logic_textio** to gain access to the I/O operations can continue to do so. The difference is that the operations will actually be provided by the **std_logic_1164** package instead.

- The package defines the strength reduction function **to_01** (see Section 8.10), to be consistent with other standard packages.

- All assertion messages produced by the package now start with the name of the package and the operation producing the message.

In addition to the new operations provided in **std_logic_1164**, the package defines a number of aliases:

- **To_std_logic_vector** and **to_slv** are defined as aliases for the conversion function **to_stdlogicvector**. Many designers have been puzzled by the absence of underscores in **to_stdlogicvector**, compared to the type name **std_logic_vector**, and find the inconsistency to be annoying. The first alias name rectifies this. The second alias name, **to_slv**, satisfies those who prefer shorter names, for example, to reduce typing.

- Similarly, **to_std_ulogic_vector** and **to_sulv** are defined as aliases for the conversion function **to_stdulogicvector**, and **to_bit_vector** and **to_bv** are defined as aliases for the conversion function **to_bitvector**.

8.2   The **Numeric_bit** and **Numeric_std** Packages

Each of the **numeric_bit** and **numeric_std** packages defines the types **unsigned** and **signed**, representing binary-coded integers as vectors of **bit** or **std_ulogic** elements, as well as operations on these types. VHDL-2008 makes the following enhancements to the packages:

- The previous version of the **numeric_std** package defined **unsigned** and **signed** as arrays of the resolved element type **std_logic**. There was no provision for unresolved elements. Since VHDL-2008 provides for resolution information to be added to elements when declaring subtypes (see Section 3.2), the **numeric_std** package revises the way the types are defined. The package defines two array types with unresolved **std_ulogic** elements, **unresolved_unsigned** and **unresolved_signed**, for use where a signal has only one source. The types **unsigned** and **signed** are now declared as subtypes with the same resolution function applied to elements as that used for **std_logic**. All of the operations in **numeric_std** are defined for the unresolved types, but can also be used with the resolved subtypes.

  In order to reduce the amount of typing required for the unresolved types, the package defines two aliases, **u_unsigned** and **u_signed**, for **unresolved_unsigned** and **unresolved_signed**, respectively.

- Both the **numeric_bit** and **numeric_std** packages define array/scalar addition and subtraction operations (see Section 4.2).

- The packages define array/scalar logic operations (see Section 4.1) and logical reduction operations (see Section 4.3).

- The packages define **maximum** and **minimum** functions, which compare the numeric values represented by the operands to determine the result. Overloaded versions are also defined with one of the parameters being of type **integer** or **natural**.

- The packages define overloaded **sla** and **sra** operators (see Section 4.8). In the case of **numeric_bit**, these operators differ from the behavior of the predefined operators on arrays of **bit** elements. Their behavior is more appropriate for vectors representing binary-coded numbers.

- The packages define two functions, **find_leftmost** and **find_rightmost**, that return the index of the leftmost and rightmost element, respectively, that has a nominated value. Comparison of elements is done using the matching equality operator. Thus, for example, **find_leftmost(V, '1')** returns the index of the leftmost occurrence of '1' or 'H' in the **unsigned** value V. If there is no such element, the function returns –1. The functions are declared as follows:

```
function find_leftmost ( arg : ArrayType; Y : std_ulogic)
                    return integer;

function find_rightmost ( arg : ArrayType; Y : std_ulogic)
                    return integer;
```

  where *ArrayType* is **unsigned** or **signed**.

- The packages define overloaded matching relational operations, "?=", "?/=", "?>", "?>=", "?<", and "?<=". These operations compare the numeric values represented by the operands. The **numeric_std** version return an 'X' result if any of the operand elements is a metalogical value (a value other than '0', '1', 'L', or 'H').

- The numeric_std package defines a complete set of strength reduction operations, in addition to the to_01 operation that was defined in the earlier version. The package also defines the is_X function. These operations are all summarized in Section 8.10.

- The packages define a complete set of string conversion functions and text I/O procedures (see Chapter 7).

- All assertion messages produced by the packages now start with the name of the package and the operation producing the message.

8.3   The Numeric Unsigned Packages

While VHDL provides the numeric_std package defining the unsigned type and associated operations, there are occasions when we would like to interpret a std_ulogic_vector value as representing a binary-coded number. Having to convert explicitly between that type and unsigned is inconvenient and clouds the intent of a model. VHDL-2008 alleviates this problem by providing the package numeric_std_unsigned. It provides the same set of operations on std_ulogic_vector values as are provided by numeric_std for unsigned values. Thus, we can perform arithmetic operations on std_ulogic_vector values without including type conversions.

VHDL-2008 also provides the numeric_bit_unsigned package. It performs an analogous purpose for bit_vector values, providing the same operations as are provided by numeric_bit for unsigned values.

8.4   The Fixed-Point Math Packages

Many digital-signal processing applications involve mathematical operations on nonintegral data. While we could use floating-point representation and hardware, that would be excessively resource-intensive in many cases. Instead, we can use a fixed-point representation, in which the radix point (analogous to the base-10 decimal point) is assumed to have a fixed position. VHDL-2008 defines a number of packages for fixed-point math that we describe in this section. The packages are all defined in the library IEEE.

For simple cases, fixed-point math amounts to integer math with scaling by a power of 2. More generally, we need to take account of rounding and overflow. The main VHDL-2008 fixed-point package, fixed_generic_pkg, has formal generic constants so that we can choose the rounding and overflow behaviors that are most appropriate for our application. The package is defined as follows:

```
package fixed_generic_pkg is
  generic (
    fixed_round_style    : fixed_round_style_type
                                    := fixed_round;
    fixed_overflow_style : fixed_overflow_style_type
                                    := fixed_saturate;
    fixed_guard_bits     : natural := 3;
    no_warning           : boolean := false
```

```
    );
  ...
```

The types fixed_round_style_type and fixed_overlow_style_type are enumeration types defined in the package fixed_float_types. The fixed_round_style generic determines the rounding behavior for operations in the package: either fixed_round, if results are to be rounded to the nearest representable value, or fixed_truncate, if results are to be truncated toward zero to the next smallest representable value. The fixed_overflow_ style generic determines the behavior on overflow: either fixed_saturate, if an overflowing result is to remain at the largest representable value, or fixed_wrap, if modulo-based behavior is required. The fixed_guard_bits generic specifies the number of extra bits of precision to use for division operations. Finally, the no_warning generic allows suppression of warning messages on conditions such as non-matching operand lengths and occurrence of metalogical values.

Since the package has generics, we must instantiate it in order to make use of it (see Section 1.2). The IEEE library includes an instance that has the default values for all of the generics. It is defined as:

```
package fixed_pkg is new IEEE.fixed_generic_pkg
  generic map (
    fixed_round_style      => IEEE.fixed_float_types.fixed_round,
    fixed_overflow_style =>
      IEEE.fixed_float_types.fixed_saturate,
    fixed_guard_bits       => 3,
    no_warning             => false
  );
```

The package fixed_generic_pkg (and any instance of it) defines types for unsigned and signed fixed-point representation in the form of vectors of std_ulogic elements. The base type for unsigned representation is unresolved_ufixed, declared as:

```
type unresolved_ufixed is array (integer range <>) of std_ulogic;
```

The name u_ufixed is defined, for convenience, as an alias to unresolved_ufixed. For signals with multiple sources, the type ufixed is defined as a subtype of unresolved_ufixed with resolved elements (see Section 3.2):

```
subtype ufixed is (resolved) unresolved_ufixed;
```

Objects of these types must have a descending (**downto**) index range. The whole-number part of the value is on the left of the vector, down to index 0, and the fractional part is on the right, starting at index –1. For example, given the following declaration of a fixed-point signal A:

```
signal A : ufixed(3 downto -3) := "0110100";
```

the whole-number part is A(3 **downto** 0), and the fractional part is A(–1 **downto** –3). The range of values represented is 0 to just less than 16 in steps of 0.125 (one eighth). The value represented by the default initial value is $0110.100_2 = 6.5_{10}$.

This example shows a number with both whole-number and fractional parts. In general, we can declare number with just a whole-number part (the right index being 0) or just a fraction part (the left index being –1). Indeed, we can declare numbers in which the radix point is completely outside the index range of the vector. For example, in the following:

```
variable X : ufixed(9 downto 2);
variable Y : ufixed(-5 downto -14);
```

X is an 8-bit vector representing values in the range 0 to 1020 in steps of 4, and Y is a 10-bit vector representing values in the range 0 to just less than 0.0625 (one sixteenth) in steps of 2^{-14}.

The base type defined in the package for signed representation is **unresolved_sfixed**, declared as:

```
type unresolved_sfixed is array (integer range <>) of std_ulogic;
```

As for the unsigned representation, there is an alias, **u_sfixed**, and a subtype with resolved elements, **sfixed**. Likewise, the index range for a signed value must be descending (**downto**), with the radix point being assumed between index 0 and index –1. The difference is that the signed type and subtypes use 2s-complement binary representation, with the leftmost bit being the sign bit. Thus, for example, the signal:

```
signal S : sfixed(3 downto -3);
```

represents values from –8 to just less than 8 in steps of 0.125.

The fixed-point math packages perform operations with full precision. This is illustrated in the following example:

```
signal A4_2 : ufixed(3 downto -2);
signal B3_3 : ufixed(2 downto -3);
signal Y5_3 : ufixed(4 downto -3);
...

Y5_3 <= A4_2 + B3_3;
```

The whole-number part of the addition result is one bit larger than the larger of the two operand whole-number parts. In this example, the operand whole-number parts are 4 bits and 3 bits, respectively, so the result's whole-number part is 5 bits. The fractional part of the result is the larger fractional part of the operands. In this example, the operands' fractional parts are 2 bits and 3 bits, respectively, so the result has a 3-bit fractional part. We summarize the operations provided by the packages and the sizes of the operation results in Section 8.8.

If we want to assign a fixed-point value to an object, one way is to use a string literal, for example:

```
signal A4 : ufixed(3 downto -3);
...
```

```
A4 <= "0110100";  -- string literal for 6.5
```

Alternatively, we can apply a conversion function, **to_ufixed** or **to_sfixed**, to an integer or real value. In this case, we need to specify the index range for the conversion result. There are two forms of conversion function. For the first form, we specify the left and right indices for the result, for example:

```
A4 <= to_ufixed(6.5, 3, -3);  -- pass indices
```

For the second form, we provide an object whose index range is used:

```
A4 <= to_ufixed(6.5, A4);  -- sized by A4
```

In this example, the only use of **A4** by the **to_ufixed** function is to read its left and right indices to determine the index range of the result. If **A4** were an **out**-mode signal or variable, reading would be legal in VHDL-2008; reading of **out**-mode objects is a change introduced in this revision of the language (see Section 6.3).

The use of a string literal in an arithmetic expression is problematic, since the index range of such a literal is ascending (**to**) and starts with **integer'low**. Fixed-point numbers must have descending index ranges. Instead we can use integer literals, real literals, and qualified string literals, as shown in the following examples:

```
subtype ufixed4_3 is ufixed(3 downto -3);
signal A4, B4 : ufixed4_3;
signal Y5      : ufixed (4 downto -3);
...

-- Y5 <= A4 + "0110100";            -- illegal,
Y5 <= A4 + ufixed4_3'("0110100");
Y5 <= A4 + 6.5;                     -- overloading with real
Y5 <= A4 + 6;                       -- overloading with integer
```

In the assignment marked "illegal," the index range of the string literal would be **integer'low** to **integer'low** + 6. The type qualification in the next assignment avoids this problem and results in a bit-string value with index bounds taken from the subtype **ufixed4_3**. We can safely apply the addition operator to this value and the operand **A4**, giving a result with index range 4 down to –3.

If we need to change the size of an expression result, we can use a **resize** function. As for the conversion functions, there are two forms, one in which we specify the left and right index values and the other in which we provide an object whose index range is used. For example, in the following accumulator assignment, since the addition result is one bit larger than the accumulator, we need to resize the result:

```
signal A4_3 : ufixed(3 downto -3);
signal Y7_3 : ufixed(6 downto -3);
...

-- Y7_3 <= Y7_3 + A4_3;  -- illegal, result too big
```

```
Y7_3 <= resize(arg              => Y7_3 + A4_3,
               size_res         => Y7_3,
               overflow_style => fixed_wrap,
               round_style      => fixed_truncate);
```

The **overflow_style** and **round_style** parameters allow us to control the way the value is processed if it cannot be represented exactly. The default values for these parameters are taken from the generics of the package. If those values are satisfactory, we can omit them in the **resize** call. This is shown in the following example, which uses the form of the function specifying left and right index values for the result:

```
Y7_3 <= resize (arg             => Y7_3 + A4_3,
                left_index   => 7,
                right_index => -3);
```

Full-precision arithmetic can lead to some unexpected results in expressions involving multiple operators. Consider, as an example, the following declarations and assignment:

```
signal A4, B4, C4, D4 : ufixed(3 downto 0);
signal Y6              : ufixed(5 downto 0);
signal Y7A, Y7B        : ufixed(6 downto 0);
...

Y6 <= (A4 + B4) + (C4 + D4);
```

The expression in the assignment is built as a balanced tree. Each of the additions **A4 + B4** and **C4 + D4** yields a 5-bit result, so the final result size is 6 bits. However, if we build the expression in a cascaded fashion, the result size is 7 bits. We can see this most clearly by explicitly parenthesizing the expression:

```
Y7A <= ((A4 + B4) + C4) + D4;
```

The addition **A4 + B4** yields a 5-bit result. This added to **C4** yields a 6-bit result, and the 6-bit result added to **D4** yields a 7-bit result. Since addition is associative, the following unparenthesized expression yields the same 7-bit result:

```
Y7B <= A4 + B4 + C4 + D4;
```

8.5 The Floating-Point Math Packages

The fixed-point math packages described in the previous section allow us to represent non-integral values with constant absolute precision over a given range. In some applications, however, we would prefer to use a floating-point representation, in which we can represent a greater dynamic range with a given number of bits, and have constant relative precision over the range. VHDL provides abstract floating-point types, including the type **real**, built into the language. However, they are defined to use IEEE 64-bit double-precision representation. That may not be the best choice for all applications. VHDL-2008 provides a set of packages for binary-coded floating-point representation and operations in which we can control the range and precision and many aspects of the way arithmetic operations are performed. Floating-point values are represented using the same principles as IEEE-standard floating-point, specified in IEEE Std 743 and IEEE Std 854, with a sign bit, an exponent field, and a fraction field. However, we can choose the field widths that are appropriate for our application.

The main floating-point math package, **float_generic_pkg**, is defined as:

```
package float_generic_pkg is
  generic (
    float_exponent_width : natural    := 8;
    float_fraction_width : natural    := 23;
    float_round_style    : round_type := round_nearest;
    float_denormalize    : boolean    := true;
    float_check_error    : boolean    := true;
    float_guard_bits     : natural    := 3;
    no_warning           : boolean    := false;
    package fixed_pkg is new IEEE.fixed_generic_pkg
                              generic map (<>)
  );
```

The package uses the generics to govern the behavior of operations. The generics **float_exponent_width** and **float_fraction_width** are used to determine the default size of results from the **to_float** conversion functions. The rounding mode for operations is specified by the generic **float_round_style**: **round_nearest**, **round_zero** (truncation), **round_inf** (round up toward infinity) and **round_neginf** (round down toward negative infinity). The enumeration type **round_type** is defined in **fixed_float_types**. Denormalized numbers are a form of floating-point numbers that represent very small values near zero. If the generic **float_denormalized** is true, operations in the package deal with denormalized values; otherwise, all numbers are treated as normalized. The generic **float_check_error** controls detection of invalid numbers and overflow, **float_guard_bit** specifies the number of extra bits of precision used within operations before the result is rounded, and **no_warning** allows suppression of warning messages. Finally, the generic **fixed_pkg** allows us to specify an instance of the fixed-point package (see Section 8.4) whose types are to be used for the conversion functions between fixed-point and floating-point types.

As for the fixed-point package, the **IEEE** library includes an instance of **float_generic_pkg** with default values for the generics. The package **float_pkg** is defined as follows:

```
package float_pkg is new IEEE.float_generic_pkg
  generic map (
    float_exponent_width => 8,
    float_fraction_width => 23,
    float_round_style    => IEEE.fixed_float_types.round_nearest,
    float_denormalize    => true,
    float_check_error    => true,
```

```
   float_guard_bits        => 3,
   no_warning              => false,
   fixed_pkg               => IEEE.fixed_pkg
   );
```

If we are using a combination of fixed-point and floating-point numbers in an application and need to instantiate the packages ourselves, we should instantiate the fixed-point package first, and then use the instance as the actual for the **fixed_pkg** generic in our instance of the floating-point package. For example, we might instantiate the fixed-point package as follows:

```
package my_fixed_pkg is new IEEE.fixed_generic_pkg
  generic map (
    fixed_round_style    => IEEE.fixed_float_types.fixed_round,
    fixed_overflow_style => IEEE.fixed_float_types.fixed_wrap,
    fixed_guard_bits     => 2,
    no_warning           => true
    );
```

and then instantiate the floating-point package:

```
package my_float_pkg is new IEEE.float_generic_pkg
  generic map (
    float_exponent_width => 6,
    float_fraction_width => 18,
    float_round_style    => round_zero,
    float_denormalize    => false,
    float_check_error    => true,
    float_guard_bits     => 2,
    no_warning           => true,
    fixed_pkg            => my_fixed_pkg
    );
```

The package **float_generic_pkg** (and each instance of the package) defines the base type for floating point numbers, **unresolved_float**. The alias **u_float** is a convenient shorthand for this type. There is also a subtype, **float**, which has resolved elements, for signals that have multiple sources. The declarations are:

```
type unresolved_float is array (integer range <>) of std_ulogic;

alias u_float is unresolved_float;

subtype float is (resolved) unresolved_float;
```

Objects of these types must have descending (**downto**) index ranges, for example:

```
signal A : float(8 downto -23)
            := "01000000110100000000000000000000";
```

The sign bit is at index **A'left** (bit 8 in this example), the exponent is indexed from **A'left** – 1 down to 0 (7 down to 0 in the example), and the fraction is indexed from –1 down to **A'right** (–1 down to –23 in the example). Unlike fixed-point numbers, floating-point numbers must have the sign, exponent, and fraction all present. The smallest floating-point representation supported by the package has a range of 3 down to –3. In practice, we would expect representations to be 16 bits or more, with at least 6 bits for the exponent and at least 10 bits for the fraction. For the sign bit, 0 is positive, and 1 is negative. The exponent field is an unsigned binary value representing the actual exponent biased by $2^{e-1} - 1$ (where e is the width of the exponent field). Thus, for the signal **A** declared above, the bias is 127. The actual fraction is normalized to the range of 1.0 to just less than 2.0. Since the bit to the left of the radix point would always be 1, it is not explicitly represented. Instead, the fraction field of a floating-point number just contains the bits to the right of the radix point, with a 1 bit implied to the left of the radix point.

We can use these properties of the representation to analyze the bit string used as the default initial value for the signal **A** above. The leftmost bit is 0, so the number is positive. The next 8 bits, **A(7 downto 0)**, are 10000001. As an unsigned number, this is 129. We subtract the bias, 127, to give an actual exponent of 2. The fraction field is 10100000000000000000000. We include the implied 1 bit to give an actual fraction of 1.101. Thus, the value represented is $+1.101_2 \times 2^2 = 1.625 \times 4 = 6.5$.

The packages declare a number of subtypes and aliases for IEEE standard floating-point representations. For IEEE Std 754 single-precision numbers, the declarations are:

```
subtype unresolved_float32 is unresolved_float(8 downto -23);

alias u_float32 is unresolved_float32;

subtype float32 is float(8 downto -23);
```

For IEEE Std 754 double-precision numbers (corresponding to **double float** in C, **float*8** in Fortran, and **real** in VHDL), the declarations are:

```
subtype unresolved_float64 is unresolved_float(11 downto -52);

alias u_float64 is unresolved_float64;

subtype float64 is float(11 downto -52);
```

For IEEE Std 854 extended-precision numbers (corresponding to **long double** in C and **float*16** in Fortran), the declarations are:

```
subtype unresolved_float128 is
        unresolved_float (15 downto -112);

alias u_float128 is unresolved_float128;

subtype float128 is float(15 downto -112);
```

The IEEE floating-point number standards reserve a number of representations for special purposes. In particular, numbers with all 0 or all 1 bits in the exponent field have the following meanings:

- Positive zero: 0 00000000 00000000000000000000000

- Negative zero: 1 00000000 00000000000000000000000

- Positive infinity: 0 11111111 00000000000000000000000

- Negative infinity: 1 11111111 00000000000000000000000

Note that there are two representations of 0, one positive and the other negative. Operations on floating-point values generally treat them as equivalent. In addition to these representations, a number with all 1 bits in the exponent field and at least one 1 bit in the fraction field (such as 1 11111111 00000000000000000000001) is called Not-a-Number, or NaN. Such values can result from otherwise illegal operations, such as division of zero by zero, or square root of -1.

Here are some further examples of floating-point numbers. First, the following is a large **float32** value (though not largest, as that is just less than 2**128).

0 11111110 00000000000000000000000
$= +1 \times 2^{254-127} \times (1.0 + 0.0)$
$= 2^{127} = 1.70141 \times 10^{38}$

Next, the following is the smallest **float32** value, without using denormals:

0 00000001 00000000000000000000000
$= +1 \times 2^{1-127} \times (1.0 + 0.0)$
$= 2^{-126} = 1.17549 \times 10^{-38}$

Finally, the following is a small **float32** value using denormals (though not the smallest):

0 00000000 10000000000000000000000
$= +1 \times 2^{1-127} \times (0.0 + 0.5)$
$= +1 \times 2^{-126} \times 0.5$
$= 2^{-127} = 5.87747 \times 10^{-39}$

For floating-point math operations, the result always has the largest of the exponent sizes and fraction sizes of the operands. Most often, the numbers are all of the same size, as in the following example:

```
signal A32, B32, Y32 : float(8 downto -23);
...

Y32 <= A32 + B32;
```

Further details of overloaded operations and result sizes are provided in the tables in later sections of this chapter.

If we want to assign a value to a floating-point object, we can either use a string literal or we can apply a **to_float** conversion function to an integer or real number. This is similar to the way in which we assign values to fixed-point objects (see Section 8.4). In the case of conversion functions, we can specify the result size either by specifying the exponent and fraction size, or by providing an object whose index range is used. These approaches are shown in the following example:

```
signal A_fp32 : float32;
...
A_fp32 <= "01000000110100000000000000000000";
A_fp32 <= to_float(6.5, 8, -32);  -- pass sizes
A_fp32 <= to_float(6.5, A_fp32);  -- size using A_fp32
```

As with fixed-point math, use of string literals in an expression is problematic, since their index ranges are ascending (**to**) and start with **integer'low**. The solution is the same, namely, using type-qualified string literals or using overloaded operations that accept integer or real operands. These are shown in the following example:

```
signal A, Y : float32;
...

-- Y <= A + "01000000110100000000000000000000";  -- illegal
Y <= A + float32'("01000000110100000000000000000000");
Y <= A + 6.5;        -- overloading with real
Y <= A + 6;          -- overloading with integer
```

8.6   The **Standard** Package

The predefined types and association operations in VHDL are defined in the package **standard**, residing in the library **std**. In practice, most tools have built-in implementations of the package, rather than interpreting the VHDL source code directly. VHDL-2008 enhances package **standard** by adding a number of new types and by extending the set of operations association with predefined types.

- The types **boolean_vector**, **integer_vector**, **real_vector**, and **time_vector** are now predefined. Each is an unconstrained type with **natural** as the index type, much like the predefined type **bit_vector** in earlier versions. The predefined operations on **boolean_vector** are the same as those defined for **bit_vector**. The predefined operations on **integer_vector** include the relational operators ("=", "/=", "<", ">", "<=", and ">=") and the concatenation operator ("&"). The predefined operations on **real_vector** and **time_vector** include the equality and inequality operators ("=" and "/=") and the concatenation operator ("&").

- The array/scalar logic operations and logical reduction operation (see Sections 4.1 and 4.3) are predefined for **bit_vector** and **boolean_vector**, since they are arrays with **bit** and **boolean** elements, respectively.

- The matching relational operators "?=", "?/=", "?>", "?>=", "?<", and "?<=" are predefined for **bit** and **boolean**. Further, the operators "?=" and "?/=" are predefined for **bit_vector** and **boolean_vector**. (See Section 4.5.)

- The condition operator "??" is predefined for **bit** (see Section 4.4).

- The operators **mod** and **rem** are predefined for **time**, since it is a physical type (see Section 4.7).

- The **maximum** and **minimum** operations are predefined for all of the predefined types (see Section 4.6).

- The functions **rising_edge** and **falling_edge** are predefined for **bit** and **boolean**. Prior to VHDL-2008, the **bit** versions of these functions were declared in the package **numeric_bit**. However, that was mainly to provide consistency with the **std_logic** versions defined in the **std_logic_1164** package. They rightly belong with the definition of the type on which they operate; hence, VHDL-2008 includes them in the package **standard**. The VHDL-2008 revision of the **numeric_bit** package redefines the operations there as aliases for the predefined versions.

- The **to_string** operations are predefined for all scalar types and for **bit_vector** (see Section 7.1). Further, the **to_bstring**, **to_ostring**, and **to_hstring** operations and associated aliases are predefined for **bit_vector**.

8.7   The Env Package

Previous version of the VHDL standard defined the language, but did not specify any means of accessing the simulation environment. VHDL-2008, as well as including the VHPI procedural interface (see Section 2.6), also includes a new environment package called **env**, resident in the library **std**. The package defines the following procedures:

```
procedure stop (status: integer);
procedure stop;

procedure finish (status: integer);
procedure finish;
```

When the procedure **stop** is called, the simulator stops and accepts further input from the user interface (if interactive) or command file (if running in batch mode). When the procedure **finish** is called, the simulator terminates; simulation cannot continue. The versions of the procedures that have the **status** parameter use the parameter value in an implementation-defined way. They might, for example, provide the value to a control script so that the script can determine what action to take next.

The **env** package also defines a function to access the resolution limit for the simulation:

```
function resolution_limit return delay_length;
```

One way in which we might use this function is to wait for simulation time to advance by one time step, as follows:

```
wait for env.resolution_limit;
```

Since the resolution limit, and hence the minimum time by which simulation advances, can vary from one simulation run to another, we cannot write a literal time value in such a wait statement. The use of the **resolution_limit** function allows us to write models that adapt to the resolution limit used in each simulation. We need to take care in using this function, however. It might be tempting to compare the return value with a given time unit, for example:

```
if env.resolution_limit > ns then    -- potentially illegal!
   ... -- do coarse-resolution actions
else
   ... -- do fine-resolution actions
end if;
```

The problem is that we are not allowed to write a time unit smaller than the resolution limit used in a simulation. If this code were simulated with a resolution limit greater than **ns**, the use of the unit name **ns** would cause an error. So the code can only succeed if the resolution limit is less than or equal to **ns**. We can avoid this problem by rewriting the example as:

```
if env.resolution_limit > 1.0E-9 sec then
   ... -- do coarse-resolution actions
else
   ... -- do fine-resolution actions
end if;
```

For resolution limits less than or equal to **ns**, the test returns false, so the "else" alternative is taken. For resolution limits greater than **ns**, the time literal **1.0E-9 sec** is truncated to zero, and so the test returns true. Thus, even though the calculation is not quite what appears, it produces the result we want.

8.8   Operator Overloading Summary

In this section, we summarize the operations defined in the standard packages. Table 8.1 summarizes the operand and result types for overloaded operations defined in the packages **std_logic_1164**, **numeric_std**, **numeric_bit**, **numeric_std_unsigned**, **numeric_bit_unsigned**, **fixed_generic_pkg**, and **float_generic_pkg**. The table does not include the predefined operators on the various types. In the table, the notation use is as follows:

- *LogicArrayType*: arrays of **std_ulogic** elements
- *NumericArrayType*: **signed**, **unsigned**, **ufixed**, **sfixed**, **float**, **bit_vector** with operations in **numeric_bit_unsigned** visible, or **std_ulogic_vector** with operations in **numeric_std_unsigned** visible
- *RealArrayType*: **ufixed**, **sfixed**, or **float**
- *ArrayElementType*: the element type of the operand array or arrays

TABLE 8.1   *Operand and result types*

Operators	Left	Right	Result
Binary and, or, nand, nor, xor, xnor	std_ulogic	std_ulogic	std_ulogic
	LogicArrayType	LogicArrayType	LogicArrayType
	LogicArrayType	std_ulogic	LogicArrayType
	std_ulogic	LogicArrayType	LogicArrayType
not		std_ulogic	std_ulogic
		LogicArrayType	LogicArrayType
Unary reduction and, or, nand, nor, xor, xnor		LogicArrayType	std_ulogic
=, /=, <, <=, >, >=	NumericArrayType	NumericArrayType	boolean
	NumericArrayType	integer	boolean
	integer	NumericArrayType	boolean
	RealArrayType	real	boolean
	real	RealArrayType	boolean
?=, ?/=, ?<, ?<=, ?>, ?>=	NumericArrayType	NumericArrayType	ArrayElementType
	NumericArrayType	integer	ArrayElementType
	integer	NumericArrayType	ArrayElementType
	RealArrayType	real	ArrayElementType
	real	RealArrayType	ArrayElementType
rol, ror, sll, srl	LogicArrayType	integer	LogicArrayType
sla, sra	NumericArrayType	integer	NumericArrayType
Binary +, −, *, /, mod, rem	NumericArrayType	NumericArrayType	NumericArrayType
	NumericArrayType	integer	NumericArrayType
	integer	NumericArrayType	NumericArrayType
	RealArrayType	real	RealArrayType
	real	RealArrayType	RealArrayType

(continues)

Table 8.1   Continued

Operators	Left	Right	Result
Binary +, −	NumericArrayType	std_ulogic	NumericArrayType
	std_ulogic	NumericArrayType	NumericArrayType
Unary −, abs		signed, sfixed, float	signed, sfixed, float
maximum, minimum	NumericArrayType	NumericArrayType	NumericArrayType
	NumericArrayType	integer	NumericArrayType
	integer	NumericArrayType	NumericArrayType
	RealArrayType	real	RealArrayType
	real	RealArrayType	RealArrayType

Table 8.2 summarizes the result size and/or index range for operations with array results. For arrays representing unsigned or signed integer values, only the size is relevant, as the leftmost bit is the most significant bit and the rightmost bit is the least significant bit. For fixed-point and floating-point values, the specific index bounds are relevant, as described in Sections 8.4 and 8.5. The notation for types is the same as that used in Table 8.1. In addition, L represents the left operand, R represents the right operand, A represents the array operand in the case where the other operand is scalar, and *Result* represents the result of the operation.

TABLE 8.2   *Result sizes and index ranges*

Operator	Result Type	Result Size and/or Range
Array/array and, or, nand, nor, xor, xnor	ArrayOfBits	Result'length = L'length = R'length Fixed, Float: Result'range = L'range
Array/scalar and, or, nand, nor, xor, xnor	ArrayOfBits	Result'length = A'length Fixed, Float: Result'range = A'range
not	ArrayOfBits	Result'length = R'length Fixed, Float: Result'range = R'range
rol, ror, sll, srl, sla, sra	ArrayOfBits	Result'length = A'length Fixed, Float: Result'range = A'range
+, −, *, /, rem, mod	float	maximum(L'left, R'left) down to minimum(L'right, R'right)

(continues)

Table 8.2   Continued

Operator	Result Type	Result Size and/or Range
Binary +, −	unsigned, signed	maximum(L'length, R'length) − 1 down to 0
	ufixed, sfixed	maximum(L'left, R'left) + 1 down to minimum(L'right, R'right)
*	unsigned, signed	L'length + R'length − 1 down to 0
	ufixed, sfixed	L'left + R'left + 1 down to L'right + R'right
/	unsigned, signed	L'length − 1 down to 0
	ufixed	L'left − R'right down to L'right − R'left − 1
	sfixed	L'left − R'right + 1 down to L'right − R'left
rem	unsigned, signed	R'length − 1 down to 0
	ufixed, sfixed	minimum(L'left, R'left) down to minimum(L'right, R'right)
mod	unsigned, signed	R'length − 1 down to 0
	ufixed	minimum(L'left, R'left) down to minimum(L'right, R'right)
	sfixed	R'left down to minimum(L'right, R'right)
Unary −, abs	signed	R'length − 1 down to 0
	sfixed	R'left + 1 down to R'right
minimum, maximum	DiscreteArrayType	Result'length = A'length Fixed, Float: Result'range = A'range
	unsigned, signed	maximum(L'length, R'length) − 1 down to 0
	ufixed, sfixed, float	minimum(L'left, R'left) down to minimum(L'right, R'right)

8.9   Conversion Function Summary

In this section, we summarize the conversion functions defined in the standard-logic and numeric packages. In order to present the information in more compact form, we have used some abbreviations for types and the packages in which the functions are defined: bv = **bit_vector**, slv = **std_logic_vector**, sulv = **std_ulogic_vector**, 1164 = **std_logic_1164**, nbu = **numeric_bit_unsigned**, nsu = **numeric_std_unsigned**, ns/b = **numeric_std** and **numeric_bit**, fixed = **fixed_generic_pkg**, and float = **float_generic_pkg**.

Table 8.3 shows the functions that convert between **bit** and **std_ulogic** scalar types, and between vectors of these types. The first parameter is the value to be converted. Those functions that convert from an abstract numeric value to a vector representation have a second parameter, **size**, to specify the result size.

TABLE 8.3   *Conversions between bit and standard-logic types*

Function	Return	Parameter 1	Parameter 2	Package
to_std_ulogic	std_ulogic	bit		1164
to_bit	bit	std_ulogic		1164
to_bv	bit_vector	sulv		1164
		natural	size	nbu
to_sulv	sulv	bv		1164
		slv		1164
		natural	size	nsu
		ufixed		fixed
		sfixed		fixed
		float		float
to_slv	slv	bv		1164
		sulv		1164
		natural	size	nsu
		ufixed		fixed
		sfixed		fixed
		float		float

Table 8.4 shows the functions that convert from the various numeric types to the **unsigned** and **signed** types defined in **numeric_std** and **numeric_bit**. The first parameter is the value to be converted, and the second parameter is either the size of the result (**size**) or a value of the result type whose size is used for the result (**size_res**).

The conversions from fixed-point representation have a third parameter, **overflow_style** (abbreviated to overflow in the table), of type **fixed_overflow_style_type**. The default value is the value of the generic **fixed_overflow_style**. The fourth parameter, **round_style** (abbreviated to round), is of type **fixed_round_style_type** and defaults to the value of the generic **fixed_round_style**.

The conversions from **float** have a third parameter, **round_style** (abbreviated to round), of type **round_type**, with the default being the value of the package generic **float_round_style**. The fourth parameter is **check_error** (abbreviated to chk_err), of type **boolean**, for controlling error checking during the conversion. The default is the value of the package generic **float_check_error**.

TABLE 8.4 *Conversion functions yielding **unsigned** and **signed** values*

Function	Return	Param 1	Param 2	Param 3	Param 4	Package
to_unsigned	unsigned	natural	size			ns/b
			size_res			
		ufixed	size	overflow	round	fixed
			size_res	overflow	round	
		float	size	round	chk_err	float
			size_res	round	chk_err	
to_signed	signed	integer	size			ns/b
			size_res			
		sfixed	size	overflow	round	fixed
			size_res	overflow	round	
		float	size	round	chk_err	float
			size_res	round	chk_err	

Table 8.5 shows the functions that convert from numeric types to the **ufixed** and **sfixed** types defined in **fixed_generic_pkg** and instances of that package. In the case of conversion functions defined in the floating-point packages, the definitions of **ufixed** and **sfixed** come from the instance of **fixed_generic_pkg** supplies as an actual generic package to the instance of **float_generic_pkg**. The first parameter of each function is the value to be converted. Following this are either two parameters, **left_index** and **right_index** (abbreviated to L_index and R_index in the table), to specify the index bounds of the result, or a single parameter, **size_res**, for a value whose index range is used for the result. For the conversions from **natural** or **unsigned** to **ufixed**, and for the conversions to **integer** or **signed** to **sfixed**, the default for **right_index** is 0. Additional parameters specify overflow and rounding modes (**overflow_style** and **round_style**), the number of guard bits to use (**guard_bits**), whether error checking is required (**check_error**), and whether operands of type float use denormalized representation (**denormalize**). The default values for the **overflow_style**, **round_style**, and **guard_bits** parameters come from the generics of the **fixed_generic_pkg** package; the default values for the **check_error** and **denormalize** parameters come from the generics of the **float_generic_pkg** package. Note that there are also versions of **to_ufixed** and **to_sfixed** with no parameters beyond the first **unsigned** or **signed** parameter. (This is not an error in the table layout!) These versions simply return the value of the parameter as a fixed-point value with no fractional part (that is, indexed from one less than the length down to 0).

TABLE 8.5 Conversion functions yielding *ufixed* and *sfixed values*

Function	Return	Param 1	Param 2	Param 3	Param 4	Param 5	Param 6	Param 7	Package
to_ufixed	ufixed	sulv	L_index	R_index					fixed
			size_res						
		unsigned							
			L_index	R_index	overflow	round			
			size_res	overflow	round				
		natural	L_index	R_index	overflow	round			
			size_res	overflow	round				
		real	L_index	R_index	overflow	round	guard		
			size_res	overflow	round	guard			
		float	L_index	R_index	overflow	round	chk_err	denorm	float
			size_res	overflow	round	chk_err	denorm		

(continues)

Table 8.5 Continued

Function	Return	Param 1	Param 2	Param 3	Param 4	Param 5	Param 6	Param 7	Package
to_sfixed	sfixed	ufixed							fixed
		sulv	L_index	R_index					
			size_res						
		signed							
			L_index	R_index	overflow	round			
			size_res	overflow	round				
		integer	L_index	R_index	overflow	round			
			size_res	overflow	round				
		real	L_index	R_index	overflow	round	guard		
			size_res	overflow	round	guard			
		float	L_index	R_index	overflow	round	chk_err	denorm	float
			size_res	overflow	round	chk_err	denorm		

Table 8.6 shows the functions that convert from numeric types to the **float** type defined in **fixed_generic_pkg** and instances of that package. Again, the definitions of **ufixed** and **sfixed** come from the instance of **fixed_generic_pkg** supplied as an actual generic package to the instance of **float_generic_pkg**. The first parameter of each function is the value to be converted. Following this are either two parameters, **exponent_width** and **fraction_width** (abbreviated to exponent and fraction in the table), to specify the sizes of the corresponding fields in the result, or a single parameter, **size_res**, for a value whose index range is used for the result. Additional parameters specify the rounding mode (**round_style**) and whether denormalized representation is used (**denormalize**). The default values for the field size, **round_style** and **denormalize** parameters come from the generics of the package.

TABLE 8.6   *Conversion functions yielding* **float** *values*

Function	Return	Param 1	Param 2	Param 3	Param 4	Param 5	Package
to_float	float	sulv	exponent	fraction			float
			size_res				
		unsigned					
			exponent	fraction	round		
			size_res	round			
		signed	exponent	fraction	round		
			size_res	round			
		ufixed	exponent	fraction	round	denorm	
			size_res	round	denorm		
		sfixed	exponent	fraction	round	denorm	
			size_res	round	denorm		
		integer	exponent	fraction	round		
			size_res	round			
		real	exponent	fraction	round	denorm	
			size_res	round	denorm		

The final group of conversion functions is shown in Table 8.7. These function convert from binary-coded vectors to abstract integer or real types. As in the preceding tables, the first parameter is the value to be converted, and subsequent parameters specify overflow and rounding modes (**overflow_style** and **round_style**), whether error checking is required (**check_error**), and whether operands of type float use denormalized representation (**denormalize**). The default values for these subsequent parameters come from the generics of the respective packages.

TABLE 8.7   *Conversion functions yielding **integer** and **real** values*

Function	Return	Param 1	Param 2	Param 3	Param 4	Package
to_integer	natural	bv				nbu
	natural	sulv				nsu
	natural	unsigned				ns/b
	integer	signed				
	natural	ufixed	overflow	round		fixed
	integer	sfixed	overflow	round		
	integer	float	round	chk_err		float
to_real	real	ufixed				fixed
		sfixed				
		float	round	chk_err	denorm	float

In addition to the declarations of the conversion functions, there are aliases for convenience and enhanced readability: the function **to_bv** has aliases **to_bitvector** and **to_bit_vector**; the function **to_sulv** has aliases **to_stdulogicvector** and **to_std_ulogic_vector**; and the function **to_slv** has aliases **to_stdlogicvector** and **to_std_logic_vector**.

For each binary-coded numeric type, there is a **resize** function, shown in Table 8.8. The versions yielding **bit_vector**, **std_ulogic_vector**, **unsigned**, and **signed** results have a parameter **new_size** to specify the result size, or a parameter **size_res** for an object whose index range is used for that of the result. The versions that yield fixed-point results have either two parameters (**left_index** and **right_index**) to specify the index bounds of the result, or a parameter (**size_res**) for an object whose index range is used for that of the result. They also have parameters to specify overflow and rounding modes (**overflow_style** and **round_style**), with default values coming from the package generics. Similarly, the versions that yield floating-point results have either two parameters to specify the field sizes for the result (**exponent_width** and **fraction_width**), and subsequent parameters specify rounding modes (**round_style**), whether error checking is required (**check_error**), and whether the operand and result use denormalized representation (**denormalize_in** and **denormalize_out**, respectively). The default values for these subsequent parameters come from the package generics.

TABLE 8.8 Resizing functions

Function	Return	Param 1	Param 2	Param 3	Param 4	Param 5	Param 6	Param 7	Package
resize	bv	bv	new_size						nbu
			size_res						
	sulv	sulv	new_size						nsu
			size_res						
	unsigned	unsigned	new_size						ns/b
			size_res						
	signed	signed	new_size						
			size_res						
	ufixed	ufixed	L_index	R_index	overflow	round			fixed
			size_res	overflow	round				
	sfixed	sfixed	L_index	R_index	overflow	round			
			size_res	overflow	round				
	float	float	exponent	fraction	round	chk_err	den_in	den_out	float
			size_res	round	chk_err	den_in	den_out		

Resizing an unsigned vector of type **bit_vector**, **std_ulogic_vector** or **unsigned** to produce a larger vector involves filling leftmost bits with '0'. Resizing these types to produce a smaller vector involves truncating the leftmost bits. For type **signed**, producing a larger vector involves filling the leftmost bits with copies of the operand's sign bit, and producing a smaller vector involves truncating the leftmost bits while retaining the sign bit.

Resizing a fixed-point value is similar. A ufixed vector is extended on the left or right by filling bits with '0'. An sfixed vector is extended on the left by replicating the sign bit and extended on the right by filling bits with '0'. Reducing the size of a fixed-point vector is more complicated, and depends on the overflow and rounding modes. If the vector is to be truncated on the right, a rounding mode of **fixed_truncate** causes the truncated bits to be discarded and the rightmost result bit to be unchanged, whereas a rounding mode of **fixed_round** causes the result to be rounded based on the values of the discarded bits and the rightmost result bit. If the vector is to be truncated to the left and the operand value is out of the representable range for the result, the value returned depends on the overflow style. For **fixed_saturate**, the largest representable value (for **ufixed** or for positive **sfixed** values) or the most negative representable value (for negative **sfixed** values) is returned. For **fixed_wrap**, the leftmost bits are simply truncated, which, in the case of **sfixed** values, may result in a change of sign.

Resizing a floating-point value is much more involved than resizing integral and fixed-point values. It involves determining the class of value represented by the operand (normal, denormal, zero, infinity, or NaN), resizing the exponent and fractional parts, rounding according to the **round_style** parameter, renormalizing or representing as a denormal if required, checking for errors, and transforming overflow to infinity.

8.10 Strength Reduction Function Summary

VHDL-2008 expands the definition of the strength reduction functions so that they are defined for the entire family of types based on **std_ulogic**. Functions of the following form are defined:

```
function to_01   (S : uType; XMAP : std_ulogic := '0')
                 return uType;

function to_X01  (S : uType) return uType;

function to_X01Z (S : uType) return uType;

function to_UX01 (S : uType) return uType;
```

The type *uType* includes **std_ulogic**, **std_ulogic_vector**, **unresolved_unsigned**, **unresolved_signed**, **unresolved_ufixed**, **unresolved_sfixed**, and **unresolved_float**. The value returned by each function for each operand element value is shown in Table 8.9. The functions **to_X01**, **to_X01Z**, and **to_UX01**, when applied to vector operands, convert each operand element according to the table to yield the corresponding result element. The **to_01** function, however, behaves differently. Provided all of the elements are '0', '1', 'L', or 'H', they are converted according to the table. However, if any element is a metalogical value (a value other than '0', '1', 'L', or 'H'), all elements of the result are set to the value of the **xmap** parameter. Thus, we can test any element of the result to determine whether there were any metalogical elements in the operand.

TABLE 8.9 *Strength reduction mappings*

Function	'U'	'X'	'0'	'1'	'Z'	'W'	'L'	'H'	'⎵'
to_01	xmap	xmap	'0'	'1'	xmap	xmap	'0'	'1'	xmap
to_X01	'X'	'X'	'0'	'1'	'X'	'X'	'0'	'1'	'X'
to_X01Z	'X'	'X'	'0'	'1'	'Z'	'X'	'0'	'1'	'X'
to_UX01	'U'	'X'	'0'	'1'	'X'	'X'	'0'	'1'	'X'

VHDL-2008 also expands the definition of the 'X' detection functions so that they are defined for the entire family of types based on **std_ulogic**. The function definitions are of the form:

function is_X (S : *uType*) **return** boolean;

The version for **std_ulogic** returns true if the operand is a metalogical value, or false otherwise. The versions for vector types return true if any element of the operand is a metalogical value, or false otherwise.

Chapter 9

Miscellaneous Changes

In this chapter, we describe various miscellaneous semantic and syntactic changes to constructs in VHDL-2008 not mentioned in earlier chapters. Many of the changes are relaxations to rules previously in VHDL, and will make some modeling tasks easier. Other changes are clarifications or corrections to minor inconsistencies in the language definition.

9.1 Referencing Generics in Generic Lists

In earlier versions of VHDL, a formal generic declared within a given generic list could not be used to declare other generics in that list. The same rule also applied to ports declared within a port list and to parameters declared in a subprogram's parameter list. In VHDL-2008, the rule is relaxed for generics. Thus, we can use one generic to declare subsequent generics in the list. Among other possibilities, this means we can use the value of a generic constant to constrain the size of a subsequent generic of an array type. This was one of the motivations behind the change.

EXAMPLE 9.1 *Individual propagation delay generics for array port elements*

Suppose we want to use generic constants to specify the propagation delays for an adder. The entity is declared with input and output ports that are arrays whose sizes are determined by a generic constant. We want to specify individual propagation delays for corresponding input and output port elements. The entity declaration is:

```
entity adder is
  generic ( width   : positive;
            Tpd_ab_s : time_vector(width - 1 downto 0) );
  port    ( a, b  : in  bit_vector(width - 1 downto 0);
            c_in  : in  bit;
            s     : out bit_vector(width - 1 downto 0);
            c_out : out bit );
end entity adder;
```

The generic constant **width** is used in the declaration of the second generic constant, **Tpd_ab_s**, to ensure that there is a matching propagation delay for each ele-

ment of the input and output ports. We can instantiate the entity in a design as
follows:

```
subtype byte is bit_vector(7 downto 0);
signal op1, op2, result : byte;
signal c_out : bit;
. . .

byte_adder : entity work.adder
  generic map ( width     => byte'length,
                Tpd_ab_s => (7 downto 1 => 120 ps,
                             0          => 80 ps) )
    port map    ( a => op1, b => op2, c_in => '0',
                  s => result, c_out => c_out );
```

In this instance, an actual value is given for **width**, and that is used to determine
the index range for **Tpd_ab_s**, as well as for the ports. Note that we don't have to
write the actual generics in this order. The values are determined for generics in the
order of their occurrence in the generic list, not the generic map. Thus, we could
have written the generic map as:

```
generic map ( Tpd_ab_s => (7 downto 1 => 120 ps,
                           0          => 80 ps),
              width     => byte'length )
```

though to do so might look a bit strange.

We saw several other examples of generics being used in the declarations of other
generics in our discussion of generic types, subprograms, and packages in Chapter 1.
Supporting these features was another motivation for making the change.

9.2 Function Return Subtype

In earlier version of VHDL, the value returned by a function had to belong to the subtype
defined as the function's return type. For example, in the following function:

```
subtype byte is bit_vector(7 downto 0);

function f ( x : byte ) return byte is
begin
  return '0' & x(6 downto 0);
end function f;
```

the result had to be a bit vector indexed from 7 down to 0. However, the result is a bit
vector indexed from 0 to 7, due to the rules for determining the index range of a concatenation result. Thus, in earlier versions of VHDL, this function would have caused an
error when executed.

In VHDL-2008, the rules for a function result returned by a return statement are relaxed. The value of the return expression is implicitly converted to the result subtype of the function. As a consequence, the above function does not produce an error. The conversion of the result value simply remaps the indices to the required range and direction.

9.3 Qualified Expression Subtype

In a change related to that described in Section 9.2, the rules for qualified expressions are relaxed. In earlier versions of VHDL, a type qualification stated the precise subtype of an expression. The value of the expression was required to be of that subtype. In VHDL-2008, the value is converted to the stated subtype. However, it still has to have the same base type as the stated subtype.

To illustrate the fine distinctions, suppose we declare a subtype, **byte**, as we did in Section 9.2, along with some variables:

```
subtype byte is bit_vector(7 downto 0);
variable x : byte;
variable y : IEEE.numeric_bit.unsigned(7 downto 0);
```

If we wanted to write an expression of the subtype **byte** comprising a '0' bit concatenated with the low-order 7 bits of **x**, we would have written the following in earlier versions of VHDL:

```
byte(bit_vector'('0' & x(6 downto 0)))
```

The qualified expression

```
byte'( '0' & x(6 downto 0) )
```

was illegal in earlier versions of VHDL, since the index range of the concatenation result is 0 to 7. A bit vector with that index range is not in the subtype of bit vectors with index ranges 7 down to 0. In VHDL-2008, the shorter qualified expression is legal, and converts the value to have an index range of 7 down to 0. Note that the following qualified expression is not legal:

```
byte'( y )
```

The value of **y** is a vector whose base type is **unsigned**, which is different from the base type of **byte**, namely, **bit_vector**. It is not correct to state that an **unsigned** value is a **bit_vector** value. Hence, a qualified expression is not appropriate; we should use a type conversion instead.

9.4 Type Conversions

In VHDL-2008, the rules for type conversions of array types are relaxed. Previously, when converting an expression of an array type to some other target array type, the elements of the expression and the target type had to be the same. Moreover, the types of the indices of the expression and the target had to be the same or had to be integer types. Under the revised VHDL-2008 rules, an array can be converted to any other array type, provided the elements can be converted to the target element type. If the target type is in fact a subtype that specifies the index bounds, then the converted expression must have the required number of elements.

To illustrate, suppose we have array types and signals declared as follows:

```
type exception_type is (int, ovf, div0, undef, trap);
type exception_vector is array (exception_type) of bit;
signal d_in, d_out: bit_vector(31 downto 0);
signal exception_reg : exception_vector;
```

In VHDL-2008, the type conversions:

```
exception_vector( d_in(4 downto 0) )
```

yields a vector of bits indexed from **int** to **trap**, with each element being the matching element of the slice of **d_in**, from left to right. Since the element types for the expression and the target type are both **bit**, conversion of the elements is trivial. In this example, the target type is also a subtype that specifies the index bounds, so the converted expression is required to have five elements.

We can also convert in the reverse direction:

```
bit_vector( exception_reg )
```

In this case, the target type is unconstrained, so the index range of the result comes from the index subtype defined for **bit_vector**, namely, **natural**. The subtype **natural** is declared to be an ascending range with a left bound of 0. This direction and left bound are used as the direction and left bound of the type-conversion result. The right bound comes from the number of elements. Thus, the result is a **bit_vector** value indexed from 0 to 4. We could assign that result to a slice of the **d_out** signal:

```
d_out(31 downto 27) <= bit_vector( exception_reg );
```

in which case matching elements are assigned left to right.

The rule that the element types of the converted expression and the target type must be convertible allows us to perform the following conversions:

```
variable i_vec : integer_vector(1 to 10);
variable r_vec : real_vector(1 to 10);
...

i_vec := integer_vector(r_vec);
r_vec := real_vector(i_vec);
```

Since we can convert between values of type **integer** and **real**, we can also convert between arrays with **integer** and **real** elements, respectively. Each element of the converted expression is converted to the element subtype of the target type.

9.5   Case Expression Subtype

VHDL allows the expression in a case statement to be of a one-dimensional character-array type, that is, a one-dimensional array whose elements include character literals. Examples of such types are **bit_vector**, **std_logic_vector**, and similar types. In earlier versions of VHDL, if we wrote such an expression in a case statement, the index range had to be locally static. In other words, we had to be able to determine the index bounds and direction at analysis time. The reason was that the case choices would be array aggregates or strings, and the analyzer needed to be able to check that they were all of the same correct length. An example showing where these rules are inconvenient is the following:

```
variable s : bit_vector(3 downto 0);
variable c : bit;
...

case c & s is
  ...
end case;
```

This would be illegal in earlier versions of VHDL, since the index range of the expression is not locally static. Instead, we would have to rewrite the example as:

```
variable s : bit_vector(3 downto 0);
variable c : bit;
subtype bv5 is bit_vector(0 to 4);
...

case bv5'(c & s) is
  ...
end case;
```

VHDL-2008 avoids this and other inconveniences by allowing the case expression not to have a locally static index range. All that is statically required is that the choices have the same length. When the case statement is executed, the value of the expression must have the same length as the choices. Thus, in VHDL-2008, we can complete the above example as follows:

```
variable s : bit_vector(3 downto 0);
variable c : bit;
...

case c & s is
  when "00000" => ...
  when "10000" => ...
  when others  => ...
end case;
```

All of the choices are of length five, so that determines the required length for the result of the concatenation.

A related change in VHDL-2008 is that an array aggregate containing **others** is allowed as a choice in a case statement, provided the index range of the case expression is locally static. For example, we can write a case statement as follows:

```
variable s : bit_vector(3 downto 0);
...

case s is
  ('0', others => '1') => ...
  ('1', others => '0') => ...
  ...
end case;
```

In this example, the index range of the expression **s** can be determined at analysis time as being 3 down to 0. That means the analyzer can use the index range for the choice values. If the analyzer cannot work out the index range for the case expression, it cannot determine the index values represented by **others** in the aggregates.

9.6   Subtypes for Port and Parameter Actuals

Earlier versions of VHDL required that, for a scalar port, the actual signal be of exactly the same subtype as the port, including having exactly the same bounds and direction. A similar restriction applied to signal parameters of subprograms. For example, given the following component declaration:

```
component counter is
  port ( count : out natural; ... );
end component counter;
```

we could not use an **integer** signal as the actual signal:

```
signal count_int : integer;
...

my_counter : counter
  port map ( count => count_int, ... );
```

This would appear to be a reasonable thing to do, since the counter output is always a non-negative integer, and the count_int signal can legally take on any integer value. However, the type **integer** has different bounds from the subtype **natural**, so the association was not allowed. The motivation for the restriction was to avoid subtype checks slowing down simulation. For example, consider the procedure:

```
procedure monitor_count ( signal count   : in natural;
                                  max_val : in natural ) is
begin
```

```
    loop
      assert count <= max_val;
      wait on count;
    end loop;
end procedure monitor_count;
```

If we supplied an **integer** signal as an actual parameter, as follows:

```
signal count_int : integer;
...

monitor_count ( count_int, 50 );   -- not legal
```

each time there was an event on the signal parameter within the procedure, we would need to check that the new value belonged to the subtype of the parameter.

VHDL-2008 changes the subtype rules for scalar ports and scalar signal parameters to partially relax the subtype requirements, in a way that makes reasonable cases legal but that avoids the need for runtime subtype checks. Under the new rules, a port of mode **out**, **inout**, or **buffer** can be connected to a signal of a different subtype, provided the signal's subtype includes at least the values in the ports subtype. So, for example, the signal could be of type **integer** and the port of the subtype **natural**, since **integer** includes all of the values in **natural**. Similarly, a port of mode **in** or **inout** can be connected to a signal of a different subtype, provided the port's subtype includes at least the values in the signal's subtype. Since ports of mode **inout** are included in both conditions, the subtypes of an **inout**-mode port and the connected signal must have the same bounds, though they no longer need to have the same direction. The same rules apply to signal parameters of subprograms (though such parameters cannot be of mode **buffer**, of course).

9.7   Static Composite Expressions

VHDL requires expressions in a number of places to be locally static, meaning that they can be evaluated during analysis. An example is an expression used as a choice in a case statement. Earlier versions of VHDL limited the forms of locally static expressions producing composite results. For example, a concatenation of two strings was not locally static, even if the strings were locally static literals.

VHDL-2008 expands the kinds of expressions that are considered to be locally static. A locally static expression can be of a composite type, provided the subtype of each of the primaries in the expression is locally static and has a locally static subtype. The list of allowed primaries is expanded to include array and record aggregates, indexed array elements, array slices, and selected record elements. The operators in the expression can be any of the predefined operators or functions, or any of the operators or functions defined in the standard packages **std_logic_1164**, **numeric_bit**, **numeric_std**, **numeric_bit_unsigned**, and **numeric_std_unsigned**.

To illustrate a consequence of these changes, given the following declarations:

```
constant unsigned_const : unsigned(5 downto 0) := "000000";
type unsigned_ROM_array is array (0 to 255) of unsigned(7 downto 0);
constant unsigned_ROM : unsigned_ROM_array := (...);
```

it is now legal to write a case statement with the following choices:

```
case unsigned_value is
  "00" & unsigned_const        => ...
  ("00" & unsigned_const) + 1 => ...
  unsigned_ROM(128)            => ...
  ...
end case;
```

9.8   Static Ranges

In many modeling scenarios, we would like to use the index range of one object to declare another object. If the declaration requires the index range to be globally static, we can sometimes run into problems, as this example illustrates:

```
entity shifter is
  port ( d : in bit_vector; ... );
end entity shift_in;

architecture rtl of shift_in is
begin
  ff_gen : for i in d'range generate
    ...
  end generate ff_gen;
end architecture rtl;
```

In earlier versions of VHDL, this was illegal. The range used to define the for-generate parameter must be a globally static range; that is, it must be determined at elaboration time. The earlier rules for globally static ranges permitted use of an attribute, provided the prefix was of a globally static subtype. In this example, the subtype of **d** is bit_vector, which is not globally static.

VHDL-2008 addresses this and similar problems by recognizing that it is not the index range of the declared subtype of **d** that we are interested in. Rather, it is the index range of **d** itself, which is determined at elaboration time from the actual signal associated with the port. VHDL-2008 revises the rules for globally static expressions and ranges to allow use of attributes that provide information about the range of a prefix, provided the prefix is one of the following:

- a signal or port

- a constant or generic constant

- a type or subtype

- a globally static function call

- a variable that is not of an access type, or a variable of an access type whose designated subtype is fully constrained

For these prefixes, we can use attributes such as 'range, 'left, 'right, 'length, and 'ascending, since they provide information about the range of the prefix that is determined at elaboration time.

9.9   Use Clauses, Types, and Operations

VHDL allows a name declared in a package to be made visible in another design unit with a use clause. We commonly write

use *package_name*.**all**;

to make all of the names declared in the package visible. Alternatively, we can list individual names to identify which of the names declared in the package become visible, for example:

use *package_name.identifier*;

According to the rules of earlier versions of VHDL, strictly, if the name listed was the name of a type declared in the package, only the type name was made visible. None of the predefined operations for the type were also made visible. Moreover, for enumeration types, none of the enumeration literals were made visible, and for physical types, none of the unit names were made visible.

VHDL-2008 extends the rules for use clauses to make using type names more useful. If we write a type or subtype name in a use clause, then as well as that name becoming visible, the following additional items declared in the package become visible:

- All of the predefined operations on the type, provided they are not hidden by overloaded version also declared in the package.

- Overloaded versions of predefined operations on the type declared in the package.

- For an enumeration type or subtype, all of the enumeration literals. This includes any character literals of the type.

- For a physical type or subtype, all of the unit names for the type.

For example, suppose we declare the following package:

package stuff_pkg **is**

    type color_type **is** (red, orange, yellow, green, blue, violet);
    subtype warm_color **is** color_type **range** red **to** yellow;

    function "<" (c1, c2 : color_type) **return** boolean;
    function pretty (c : color_type) **return** boolean;

    type resistance **is range** 0 **to** 1E9 **units**

```
        Ohm;
        kOhm = 1000 Ohm;
        MOhm = 1000 kOhm;
    end units;

    subtype weak_logic is
        IEEE.std_logic_1164.std_logic range 'W' to 'H';

end package stuff_pkg;
```

Then the use clause

```
use stuff_pkg.color_type;
```

makes not only the type **color_type** visible, but also the enumeration literals **red** through **violet**, the predefined operations on **color_type** other than "<", and the overloaded "<" operator declared in the package. It does not make the function **pretty** visible, since it is not an overloaded version of a predefined operation. If we write the use clause

```
use stuff_pkg.warm_color;
```

it makes the subtype **warm_color** visible, along with all of the enumeration literals for **color_type** (not just those in the subtype) and the operations for **color_type**.

If we write the use clause

```
use stuff_pkg.resistance;
```

it makes the type **resistance** visible, along with the unit names **Ohm**, **kOhm**, and **MOhm**, and the predefined operations on **resistance**.

Finally, if we write the use clause

```
use stuff_pkg.weak_logic;
```

all we get is the subtype name **weak_logic** made visible, since none of the enumeration literals or operations are declared in the package **stuff_pkg**.

9.10 Hiding of Implicit Operations

Earlier versions of VHDL had a subtle interaction between the rules for use clauses and the rules for overloading predefined operations. This interaction caused a problem during development of the **numeric_bit_unsigned** and **numeric_std_unsigned** packages. When we declare a type in a package, the predefined operators for the type are implicitly declared in that package. For example, the type **std_logic_vector** is declared in package **std_logic_1164**, and so the predefined operators such as "<" are also declared implicitly in **std_logic_1164**. If we were to declare an overloaded version of "<" for **std_logic_vector** in the same package, it would hide the predefined version. A use clause for the package with the overloaded operator would make the overloaded version visible, not the predefined version.

On the other hand, we can write a separate package that defines overloaded operators. The package **numeric_std_unsigned** does just that for the **std_logic_vector** type. It declared overloaded comparison operators that perform arithmetic comparisons, whereas the predefined comparison operators in **std_logic_1164** perform lexicographic comparisons.

Now suppose we write a design unit that uses both **std_logic_1164** and **numeric_std_unsigned**:

```
library IEEE;
use IEEE.std_logic_1164.all, IEEE.numeric_std_unsigned.all;
...
```

The question is, which version of the "<" is visible in the design unit? Since the two versions are declared in different packages, under the old VHDL rules, the explicitly declared version did not hide the implicitly declared version. Since the use clauses made two version potentially visible, neither version was made visible. VHDL-2008 revises the rules for use clauses so that an overloaded version of a predefined operation explicitly declared in one package, when used by a use clause, hides the use of an implicitly declared version from another package. So in the above example, the "<" operator explicitly declared in **numeric_std_unsigned** is made visible, and the implicit version from **std_logic_1164** remains hidden.

9.11 Multidimensional Array Alias

VHDL allows us to write an alias for an object, and to specify a subtype with which to view the object. For example, we can write:

```
signal s : bit_vector(15 downto 0);
alias bigendian_s : bit_vector(0 to 15) is s;
```

In earlier versions of VHDL, the rules for aliases of arrays were unclear, and prohibited aliases of multidimensional array objects. Thus, the following was illegal in earlier versions of VHDL:

```
type bit_matrix is
  array (natural range <>, natural range <>) of bit;
signal s : bit_matrix(15 downto 0, 15 downto 0);
alias bigendian_s : bit_matrix(0 to 15, 0 to 15) is s;
```

VHDL-2008 clarifies the rules and removes the restriction, making the above example legal.

9.12 Others in Aggregates

VHDL allows an array aggregate to include an **others** choice, provided the index range of the array can be determined from the context. We need the index range in order to work out what index values are implied by the **others** choice. In earlier versions of

VHDL, the rules identifying where an array aggregate could include **others** omitted a number of useful cases. VHDL-2008 rectifies the omissions.

One place where earlier versions of VHDL did not allow us to use **others** in an array aggregate was as a default value for a generic constant declared to be of a constrained array type. For example, the aggregate in the following declaration would be illegal in earlier versions of VHDL:

```
entity adder32 is
  generic ( Tpd : time_vector(31 downto 0)
                    := (others => 100 ps) );
  port    ( a, b : in  bit_vector(31 downto 0);
          ( s    : out bit_vector(31 downto 0) );
  end entity adder32;
```

While we could certainly write out an aggregate with 32 elements, or rewrite the aggregate as:

```
(31 downto 0 => 100 ps)
```

this would be less convenient. VHDL-2008 rectifies the omission, making the aggregate in the generic list above legal. Moreover, when combined with the change described in Section 9.1, the following becomes legal in VHDL-2008:

```
entity adder is
  generic ( width : positive;
            Tpd : time_vector(width-1 downto 0)
                    := (others => 100 ps) );
  port    ( a, b : in  bit_vector(width-1 downto 0);
          ( s    : out bit_vector(width-1 downto 0) );
  end entity adder;
```

Other places where aggregates with **others** were previously omitted but are now allowed are:

- As an actual expression in a port map associated with a port (or an element of a port) that is of a fully constrained array subtype.

- As an actual expression associated with a slice of a parameter in a subprogram call, or with a slice of a generic in a generic map, or with a slice of a port in a port map.

- As the right-hand side of an assignment statement where the assignment target is an element of an object and the element is of a fully constrained array type.

- As the right-hand side of an assignment statement where the assignment target is a slice of an object.

9.13 Attribute Specifications in Package Bodies

Earlier versions of VHDL did not allow attribute declarations or specification in package bodies. This omission is rectified in VHDL-2008. As a consequence, the following is now legal, whereas it was illegal in previous versions:

```
package utility_pkg is
  type lookup_ROM is array(0 to 15) of bit_vector(7 downto 0);
  constant lookup : lookup_ROM;
end package utility_pkg;

package body utility_pkg is

  function get_cpu_time return delay_length;
  attribute foreign of get_cpu_time : function is
    "VHPIDIRECT libutility get_cpu_time";
  ...

  constant lookup : lookup_ROM := ( ... );

  attribute logic_block : boolean;
  attribute logic_block of lookup : constant is true;

end package utilty_pkg;
```

The package declaration contains an attribute specification to decorate the function **get_cpu_time** with the **'foreign** attribute. This allows the function, declared privately within the package body, to have a VHPI implementation. The package body also declares the synthesis attribute **'logic_block** and decorates the deferred constant **lookup** with the attribute. These attribute declarations and specifications are all now legal in VHDL-2008.

9.14 Attribute Specification for Overloaded Subprograms

The way attribute specifications for overloaded subprograms were defined in earlier versions of VHDL led to some undesirable consequences in some cases. One of the rules was that, if we wrote an attribute specification for a name, and there were multiple overloaded subprograms of that name, then all of the overloaded versions were decorated. The problem arose if a given name was used for both procedures and functions. For example, suppose we declared the following subprograms and tried to decorate them:

```
procedure add ( a, b : in integer; s : out integer );
procedure add ( a, b : in real; s : out real );
function add ( a, b : integer ) return integer;
function add ( a, b : real ) return real;

attribute built_in : boolean;
```

```
atribute built_in of add : procedure is true;
attribute built_in of add : function is false;
```

The problem was that, in earlier versions of VHDL, the subprogram identified by the names did not take account of the subprogram kind in the attribute specification. That caused us to run foul of the rule that all of the items identified by the name had to be of the specified kind. Thus, in the example, the name **add** in both attribute specifications identifies all four overloaded versions. Even though two are not procedures for the first specification, and two are not functions for the second specification. We could avoid this problem by writing separate attribute specifications for each subprogram, using a signature to differentiate them, but that would require four attribute specifications instead of two:

```
atribute built_in of
  add[integer, integer, integer] : procedure is true;
atribute built_in of
  add[real, real, real] : procedure is true;
attribute built_in of
  add[integer, integer return integer] : function is false;
attribute built_in of
  add[real, real return real] : function is false;
```

Clearly, this is more cumbersome. VHDL-2008 tidies up the rules for attribute specifications by taking account of the kind of item. If the name identifies multiple overloaded subprograms, only those of the specified kind are actually decorated. Thus, we can legally write the original two attribute specifications above. Only the two procedures are decorated with the attribute value **true**, and only the two functions are decorated with the attribute value **false**.

9.15 Integer Expressions in Range Bounds

In certain cases in earlier versions of VHDL, we could not use a numeric literal as one bound for a range and a more complex expression as the other bound. An example is a for loop written as follows:

```
for i in 0 to 2**N - 1 loop ...
```

where N is a constant of type integer. A similar problem could arise in a for generate statement and in an index range for an array type definition.

The underlying reason, in earlier versions of VHDL, was that the bounds are both of the predefined type **universal_integer**, but VHDL only implicitly converted the bounds from **universal_integer** values to **integer** if both were simple expressions consisting just of a numeric literal or an attribute value. If that was not the case, the bounds were required to be of a type other than **universal_integer**. In the example above, the fact that the right bound is a more complex expression prevented the implicit conversion being performed, leaving both bounds of the type **universal_integer**, which was illegal.

VHDL-2008 rectifies the anomaly by removing the requirement that both bounds be simple expressions for the implicit conversion to **integer** to apply. Instead, the implicit conversion is done if both bounds are of type **universal_integer**, regardless of the complexity of the expressions. Thus, in the above example in VHDL-2008, the conversion is performed, leading to the loop parameter i being of type **integer**.

9.16 Action on Assertion Violations

Assertion statements and report statements allow specification of the severity level (**note**, **warning**, **error**, or **failure**) associated with a message. Earlier versions of VHDL did not specify the effect of different severity levels, beyond saying that it should be included in the message. Different simulators have taken different approaches. Most will stop execution on some given severity level or greater, but they differ in the threshold at which they stop. Some stop on **error** or **failure**, whereas others stop only on **failure**. This has caused portability problems for users writing simulation control scripts and for developers of packages that include assertions.

VHDL-2008 recommends that a simulator continue execution for an assertion or report statement with severity level of **error** or less. This should help reduce the portability problems, as tool implementers converge upon the recommendation.

9.17 'Path_Name and 'Instance_Name

VHDL includes two predefined attributes, **'path_name** and **'instance_name**, that give string values representing the path through the design hierarchy from the root to the prefix item. These attributes were introduced in VHDL-93.

One omission from the rules for forming the attribute values in earlier versions of VHDL was provision for overloaded operators. Both attributes included the name of a subprogram enclosing an item, but only allowed for an identifier to represent the subprogram name. An overloaded operator, on the other hand, has an operator symbol in quotation marks as its name. VHDL-2008 amends the rules to allow for this scenario. Thus, given the following design hierarchy:

```
entity e is
end entity e;

architecture a of e is
begin
  proc : process is
    type T is ...
    function "+" ( a, b : T ) return T is
      variable s : integer;
    begin
      ...
    end function "+";
  begin
    ...
```

```
    end process proc;
  end architecture a;
```

the value of s'pathname is the string

```
  :e:proc:"+":s
```

and the value of s'instance name is the string

```
  :e(a):proc:"+":s
```

In VHDL-2000, protected types for shared variables were introduced. A protected type encapsulates declarations and is instantiated when a shared variable is elaborated. Thus, a shared variable constitutes part of the hierarchy of a design. The path to an item declared within a protected type descends through a shared variable. Nonetheless, neither VHDL-2000 nor VHDL-2002 made provision for a shared variable's name in rules for the 'path_name and 'instance_name attributes. VHDL-2008 rectifies this omission. To illustrate, suppose we declare a protected type in a package as follows:

```
package sharing_pkg is
  type int_mailbox is protected
    . . .
  end protected int_mailbox;
end package sharing_pkg;

package body sharing_pkg is
  type int_mailbox is protected body
    variable int : integer := 0;
    . . .
  end protected body int_mailbox;
end package body sharing_pkg;
```

We then declare a shared variable of this type within an architecture:

```
entity e is
end entity e;

architecture a of e is
  shared variable v : work.sharing_pkg.int_mailbox;
begin
  . . .
end architecture a;
```

The value of int'path_name for the encapsulated variable within the instance v is:

```
:e:v:int
```

Similarly, the value of int'instance_name for that encapsulated variable is:

```
:e(a):v:int
```

Note that in both cases, the path does not go through the package in which the protected type is declared. Rather, the path follows the instantiation hierarchy through the shared variable.

9.18 Non-Nesting of Architecture Region

In versions of VHDL prior to 2002, there were some anomalies relating to the way in which entity and architecture names were defined. A literal interpretation of the rules meant that entity and architecture names could not be referenced. Since this was clearly not intended, different implementations of VHDL made different interpretations, resulting in some incompatibilities between tools.

In those earlier versions, the declarative part of an architecture was intended to be an extension of the declarative part of the corresponding entity. However, it was not clear where the architecture name itself was declared. Problems became apparent when the architecture name was the same as the entity name, a common practice for many designers. Under some interpretations of the rules for scope and visibility of names, the architecture name hid the entity name, whereas under other interpretations, making the entity and architecture names the same was illegal.

VHDL-2002 sought to clarify the rules by specifying that an architecture was no longer to be considered as an extension of the entity's declarative part. Instead, the architecture formed a nested declarative region, logically positioned within and at the end of the entity declarative part. Thus, the architecture name was officially declared within the entity, and could hide the entity name if it was the same. This approach very neatly solved the naming problems with minimal change to the scope and visibility rules. However, the nesting structure could become apparent in a number of ways, and turned out not to be what users wanted. In practice, VHDL implementers did not revise their tools to conform with the new rules, and so the old problems remained.

VHDL-2008 reverts the rules for architectures to specify that an architecture is logically an extension of the corresponding entity declarative part. Moreover, the scope and visibility rules are revised to clarify the issues that led to different interpretations in earlier versions of VHDL. In particular, an architecture name is no longer considered to be declared within the entity declaration. Rather, an architecture name is given special consideration, and the places where it makes sense to refer to it are explicitly listed. As implementations converge on the new rules, the inconsistencies and incompatibilities from earlier versions of VHDL should disappear.

9.19 Purity of Now

VHDL-93 introduced the distinction between pure and impure functions. Essentially, a pure function returns the same result when called with the same parameters, whereas an impure function, by virtue of being allowed to reference items outside its declaration, may return different results.

The standard function **now**, defined in package **standard**, has no parameters and returns the current simulation time. Clearly, since the simulation time can change from one call to another, the function should be impure, and in VHDL-93 it was. In VHDL-

2002, its declaration was changed to pure in order to address some concerns in another standard related to VHDL. However, this caused more problems than it solved, particularly since that other standard was revised to avoid the concerns. VHDL-2008 rectifies the anomaly by changing the declaration of **now** back to an impure function.

9.20 Delimited Comments

Earlier versions of VHDL have single-line comments, starting with the characters "--" and extending to the end of the line. VHDL-2008 keeps this style of comment, but also adds *delimited comments*, starting with the characters "/*" and extending to the closing characters "*/". The opening and closing characters can be on different lines, or can be on the same line. Moreover, there can be further VHDL code on the line after the closing characters. Some examples are:

```
/* This is a comment header that describes
   the purpose of the design unit. It contains
   all you ever wanted to know, plus more.
*/

library IEEE; context IEEE.IEEE_STD_CONTEXT;
entity thingumy is
  port ( clk   : in std_logic; -- keeps it going
         reset : in std_logic  /* start over */
         /* other ports to be added later */ );
end entity thingumy;
```

Since the text in comments is ignored, it may contain comment delimiters. Mixing comment styles can be quite useful. For example, if we use delimited comments in a section of code, and we want to "comment out" the section, we can use single-line comments:

```
-- This section commented out because it doesn't work
-- /* Process to do a complicated computation
--    involving lots of digital signal processing.
-- */
-- dsp_stuff : process is
-- begin
--   assert 2 + 2 = 4; -- make sure we're in the right universe
--   ...
-- end process dsp_stuff;
```

However, we should be aware that comments do not nest. For example, the following is ill-formed:

```
-- Here is the start of the comment: /* A comment extending
                                         over two lines */
```

The opening "/*" characters occur in a single-line comment, and so are ignored. Similarly, we cannot reliably use delimited comments to comment out a section of code, since the section might already contain a delimited comment:

```
/* Comment out the following code:
signal count : unsigned(5 downto 0); /* event counter */
*/
```

In this case, the occurrence of the characters "*/" on the second line closes the comment started on the first line, making the orphaned delimiter "*/" on the third line illegal. Provided we avoid pitfalls such as these, delimited comments are a useful addition to the language.

9.21 Tool Directives

In Section 2.5, we mentioned that protect directives for IP protection are a form of tool directives. VHDL-2008 adds tool directives as a way of embedding information for use by tools in a VHDL model. A tool directive takes the form:

```
` identifier ...
```

The grave accent character (sometimes called a "back-tick") is followed by an identifier that determines the action to be performed or the kind of information provided. The rest of the text on the line, if any, provides any further information required. The directive finishes with the end of the line. VHDL-2008 defines protect directives, with the identifier **protect**, but does not define any other kinds of directives. Implementations may define their own kinds of directives and place requirements on the text that follows the identifier.

9.22 New Reserved Words

As new features are added to VHDL, new reserved words are usually required. In VHDL-2008, the new reserved words added are

context

>Used in context declarations and context references (see Section 2.3).

default

>Can be used in the generic map of a formal generic package (see Section 1.6).

force

>Used in a simple forcing assignment (see Section 2.2) and in conditional and selected forcing assignments (see Section 5.1.2).

parameter

Optionally precedes the parameter list in a subprogram (see Section 1.4).

release

Used in a release assignment (see Section 2.2).

In addition, the following PSL reserved words are also VHDL-2008 reserved words

assert
assume
assume_guarantee
cover
default
fairness
property
restrict
restrict_guarantee
sequence
strong
vmode
vprop
vunit

Note that **assert** was previously a reserved word, but now gains another use. Also, **default** has two uses, one in a PSL default clock declaration, and the other in the generic map of a formal generic package.

Since a reserved word cannot be used as an identifier, there is potential for backward incompatibility. If a model written in an earlier version of VHDL uses one of the new reserved words as an identifier, it is not legal VHDL-2008. However, most tools provide options to analyze a model using the rules of earlier versions of the language and allow a design to be composed of design units written in a mixture of language versions. So, in practice, backward incompatibility is not an insurmountable problem.

9.23 Replacement Characters

Earlier versions of VHDL allowed certain characters in models to be replaced with others. Specifically,

- A vertical bar ("|") could be replaced by an exclamation mark ("!").

- The number-sign characters ("#") in a based numeric literal could be replaced by colon characters (":").

- The double-quote characters (""") in a string literal could be replaced by percent characters ("%").

Most VHDL users are unaware that such replacements were permitted. Nonetheless, they were allowed in the initial version of the language, since at that time, some computers used the EBCDIC character code, which did not include the replaced characters. In VHDL-2002, replacement characters were included in a list of deprecated features. The axe falls in VHDL-2008.

Chapter 10

What's Next

In this final chapter, we look into the near future and give a preview of some features that are being developed by the Accellera VHDL Technical Committee (the VHDL-TC). The focus of the new features is verification and system-level modeling. From a language perspective, VHDL already provides some support for these modeling tasks, in the form of records, access types (pointers), and protected types (shared variables). The new features being developed include class types, verification data structures, randomization, and functional coverage. As much as possible, the new features will build on existing features in the language. Where no existing features meet a need, the new features added will be designed to integrate with the syntax and semantics of other existing features.

The language features described in this chapter are currently being actively worked on in committee. There is a great deal more to each of the proposals than is presented here. However, since the details are still subject to revision, it is too early to publish them. We present an overview to whet the appetite and encourage participation in the development process. The long-term plan is to evolve VHDL from a Hardware Description Language to a Verification and Hardware Description Language.

10.1 Object-Oriented Class Types

Class types are a foundation feature for both data structures and randomization. Briefly, a class type encapsulates data and provides operations, called methods, to access and update the data. An object is an instance of a class type. A subclass can inherit data and operations from a superclass, and in doing so, can override the implementation of inherited operations. References to objects can be polymorphic, meaning that they can refer to an object of a nominated class or of any subclass. When a method is invoked via a polymorphic reference, the overriding method of the referenced object's specific class is executed. This is referred to as dynamic dispatch.

The object-oriented features under development involve basing class type on protected types, since the latter already provide encapsulation of data and methods. We can view a protected type as a class type without inheritance or dynamic dispatch. A side benefit of this approach is that it simplifies the prototyping of data structures and communication protocols using the existing language. For example, if we are to use object-oriented features for verification, we need to implement a number of data structures, including lists, FIFOs, mailboxes, transaction interfaces, scoreboards, and memories. We can already build many of these data structures using protected types. However, the class extension increases their parameterizability, and hence, reusability.

EXAMPLE 10.1   *Using protected class types for FIFO communication*

A protected class type definition for a simple bounded FIFO is similar to a protected type as currently provided in VHDL, requiring a declaration and a body:

```
type BoundedFIFO is protected class
  procedure put ( e : in  element_type );      -- methods
  procedure get ( e : out element_type );
end protected class BoundedFIFO;

type BoundedFIFO is protected class body

  constant size : positive := 20;
  type element_array is array (0 to size-1) of element_type;

  variable elements : element_array;
  variable head, tail : natural range 0 to size-1 := 0;
  variable count : natural range 0 to size := 0;

  procedure put ( e : in element_type ) is
  begin
    if count = size then wait until count < size; end if;
    elements(head) := e;
    head := (head + 1) mod size;  count := count + 1;
  end procedure put;

  procedure get ( e : out element_type ) is ...

end protected class body BoundedFIFO;
```

The class declaration contains the publicly visible data members and methods, in this case, just the methods **put** and **get**. The body contains the private data members and the implementations of the methods. The implementation of the **put** method in this example illustrates a new form of conditional wait that allows a method to suspend until a condition becomes true. This is an enhanced form of concurrency control included in the proposal.

Among the other features included in the class proposal, beyond the basic object-oriented features, is provision for objects of class types to be included as subprogram parameters and ports of components and entities. Moreover, the proposal adopts Java-like interface definitions and multiple inheritance. An interface (as in the Java sense) specifies methods that a class must implement. A given class may implement more than one interface, as well as inheriting from a superclass. The interface feature provides a very powerful abstraction mechanism, as the following example illustrates.

EXAMPLE 10.2   *Communication interfaces*

A modular design involves a number of components that communicate with one another. Communication involves a producer putting data into a communications data structure and a consumer getting data from the data structure. We can express the notions of putting and getting as two distinct interfaces that can be implemented by a wide variety of different data structures. The details of an implementation are not relevant to a producer or consumer, only the signature of each interface. Thus, we define two interfaces as follows:

```
type putable is interface
  procedure put     (e : in element_type);
  procedure try_put (e : in element_type;  ok : out boolean);
end interface putable;

type getable is interface
  procedure get     (e : out element_type);
  procedure try_get (e : out element_type; ok : out boolean);
end interface getable;
```

A mailbox class might implement the **putable** and **getable** interfaces as follows:

```
type mailboxPCType is protected class implements putable, getable
  impure function flag_up return boolean;
  procedure put     (e : in element_type);
  procedure try_put (e : in element_type;  ok : out boolean);
  procedure get     (e : out element_type);
  procedure try_get (e : out element_type; ok : out boolean);
end protected class mailboxPCType;
```

The class body would include implementations of the **flag_up** function and the four procedures.

For a parameter of a subprogram or a port of a component or entity, we can specify an interface as the type rather than a specific concrete class type, as shown in the components declared in the following model:

```
entity tlm is
end entity tlm;

architecture structural of tlm is

  component producer is
    port ( shared variable data_source : inout putable );
  end component producer;
```

```
component consumer is
  port ( shared variable data_sink : inout getable );
end component consumer;

shared variable MailBox : mailboxPCType;

begin

  u_producer : producer port map ( data_source => MailBox );
  u_consumer : consumer port map ( data_sink   => MailBox );

end architecture tlm;
```

For the producer and consumer components, we could bind any entities that communicate through the **putable** and **getable** interfaces, respectively. Moreover, we could use any concrete class that implements the interfaces, such as the **mailboxPC-Type** class, to connect the components.

10.1.1 Standard Components Library

Object-oriented language features allow factoring of common code into abstract super-classes, with the inheriting subclasses refining behavior by adding data members and overriding methods. These aspects are put to good use in other object-oriented languages to provide suites of reusable data structures. Examples are the Standard Template Library in C++, the Java collections library, and the Booch Components in Ada.

The VHDL-TC is planning to develop a standard components library along a similar vein. It will include packages defining class types for data structures, using formal generic types for the contained elements. The library will also include packages defining reusable verification components, such as communication interfaces and classes and support for stimulus generators and checkers. The plan is, ultimately, to provide components similar in nature to those that support the Advanced Verification Methodology (AVM) and the Verification Methodology Manual (VMM).

10.2 Randomization

One problem in verifying a design with a large verification space and numerous configurable features is how to write enough test cases to adequately test all of the features. For some designs, using the algorithmic features of the VHDL is sufficient to generate the test cases. For other designs, randomization is more appropriate. The randomization proposal under consideration by the VHDL-TC introduces three forms of randomization: basic randomization, class-based randomization, and procedural randomization. These features are similar to those of SystemVerilog; however, their syntax is consistent with other VHDL constructs.

The intent of basic randomization is to provide individual random values. Basic randomization is implemented in a predefined class providing function methods **RandReal**, **RandInt**, **RandSlv**, **RandUnsigned**, and **RandSigned**. Since these functions are encapsulated in a class, the seed is also stored in the class and does not need to be passed as a parameter in a procedure call. Each of the functions has parameters that allow the result to be scaled to a particular range of values.

In class-based randomization, related objects are included as data members of a class. Within the class, relationships are written between the class members. During randomization (using a built-in method), class members are randomized taking these relationships into account. In this manner, meaningful values can be generated to specify a transaction or a sequence of transactions.

EXAMPLE 10.3   *Randomized bus traffic*

A class to generate random bursts of traffic on a bus is shown below. In this class, BurstLen specifies the number of values to generate, and BurstDelay specifies the number of cycles to insert between bursts. BurstLen is randomized with values between 1 and 10, inclusive, and BurstDelay is randomized as a function of BurstLen. If BurstLen is less than 3, BurstDelay has a value between 1 and 6, inclusive; otherwise, BurstDelay has a value between 3 and 10, inclusive.

```
type TxPacketCType is class

  rand variable BurstLen   : integer;    -- Public Variables
  rand variable BurstDelay : integer;

  constraint BurstPkt is (
    BurstLen in (1 to 10);
    BurstDelay in (1 to 6) when BurstLen <= 3 else
    BurstDelay in (3 to 10);
  );

end class TxPacketCType;
```

A combination of basic and class-based randomization is shown below. The call to **randomize** generates a random value for both BurstLen and BurstDelay. Since these objects are public class members, their value can be accessed directly using the same "." notation that is used for records.

```
TxProc : process
  variable TxPacket : TxPacketCTType;
  variable RV : RandClass;
begin
  ...
  TxOuterLoop: loop

    TxPacket.randomize;

    for i in 1 to TxPacket.BurstLen loop
      DataSent := RV.RandSlv(0, 255, DataSent'length);
```

```
        Scoreboard.PutExpectedData(DataSent);
        WriteToFifo(DataSent);
      end loop;

      wait for TxPacket.BurstDelay * tperiod_Clk - tpd;
      wait until Clk = '1';

    end loop TxOuterLoop;
    ...
  end process TxProc;
```

Procedural randomization is a proposed enhancement to VHDL's sequential constructs. A RandCase feature provides random choice among statements, and a sequence feature allows randomly or deterministically ordered sequences of statements to be executed.

EXAMPLE 10.4 *Random permutation*

We can use the RandCase feature to choose a random permutation of statements to execute. The code below loops three times, randomly selecting a statement to execute in each iteration. The RandCase feature uses weights to express relative frequencies of choice among the alternatives. After a statement has been executed, the weight for that alternative is set to 0, preventing the alternative from being executed a second time.

```
    I0 := 1; I1 := 1; I2 := 1;
    for i in 1 to 3 loop

      randcase is
        with I0 =>
          CpuWrite(CpuRec, DMA_WORD_COUNT, DmaWcIn);
          I0 := 0; -- modify weight

        with I1 =>
          CpuWrite(CpuRec, DMA_ADDR_HI, DmaAddrHiIn);
          I1 := 0; -- modify weight

        with I2 =>
          CpuWrite(CpuRec, DMA_ADDR_LO, DmaAddrLoIn);
          I2 := 0; -- modify weight

      end randcase;

    end loop;
    CpuWrite(CpuRec, DMA_CTRL, START_DMA or DmaCycle);
```

10.3 Functional Coverage

Functional coverage is intended to supplement other forms of coverage. Tool based code coverage provides information about what parts of a design are exercised during a simulation. However, it cannot test whether an aspect of the specification for the design is actually implemented. Functional coverage features, on the other hand, allow us to measure the occurrence of difference categories of data values during a simulation. We can thus determine whether processing of categories of interest has been exercised. To measure functional coverage, we specify a bin (a value or range of values) for each category of a data object. During simulation, for each bin, the tool records the number of transactions that produce values in the bin. We can analyze the result to identify bins for which no transactions occurred, and adjust our stimulus generation or randomization constraints accordingly.

The VHDL-TC plans to incorporate functional coverage features into a future extension of VHDL. The details of language features are yet to be determined.

10.4 Alternatives

One question that comes up frequently is, why update VHDL? Instead, why not adopt SystemVerilog as the verification language? The answer is much simpler than one would expect. From a language perspective, VHDL already includes many system-level modeling features, such as records, access types (pointers), and protected types. Many of these features can be enhanced with relatively little impact on the language, and new features can be added in a way that integrates cleanly with existing features. From a project perspective, organizations using VHDL already have significant experience using the language. If a verification engineer is needed for a project, the organization has a pool of people familiar with VHDL. A person from that pool can build on their existing knowledge of VHDL, provided the language includes the necessary verification features. If they were to adopt SystemVerilog, they would not only have to learn a language that is quite idiomatically different, but they would also have to manage a multilingual design/verification environment. Both of these issues would adversely affect their productivity.

10.5 Getting Involved

Standards development is a volunteer-run effort, and depends on your participation. As you become an experienced VHDL design and/or verification engineer, it is both your right and responsibility to participate. You can participate by submitting enhancement requests, participating in the standards groups, helping with funding, and helping with vendor support.

One person can make a difference. No matter how hard the VHDL-TC works, without your ideas, the group may overlook the changes you desire. You can submit your enhancement requests using the web page at http://www.eda.org/vasg.

VHDL standards are co-developed by IEEE and Accellera. Currently most of the new technical development is done by the Accellera VHDL-TC. The IEEE VHDL Analysis and Standardization Group (VASG) resolves issues with the current IEEE VHDL standard and conducts balloting for new IEEE versions of the standard. For more information about the Accellera VHDL-TC, see http://www.accellera.org/vhdl, and for more information about the IEEE VASG, see http://www.eda.org/vasg.

Volunteers run these standards groups, and members tend to work on what interests them personally. For a request to become a proposal and then a language feature, someone has to champion it. The best way to make this happen is to participate. Participation is open to anyone who has the background and is willing to invest the time. You can join the technical subcommittees, participate in email reflectors, attend teleconferences, attend in-person meetings, and actively participate in all discussions. Most technical decisions are held at a level where everyone can contribute. When decisions have conflicting choices, the issue is put to a member vote. To have a member vote in the Accellera VHDL-TC, your company must join Accellera. To have a member vote in the IEEE VASG, you need to join the parent group, IEEE Design Automation Standards Committee (DASC, see http://www.dasc.org) and maintain an active history of voting participation.

While much of the work is volunteer based, the task of integrating the language change proposals and editing the standard is a time-intensive task and is undertaken by a paid technical editor. This person is a VHDL expert with deep language design knowledge. Currently, this position is funded through Accellera. If your company is able, please encourage them to become an Accellera member and help fund future revisions of the VHDL standard.

Finally, ongoing evolution of VHDL requires vendor support. Part of achieving this is to understand why vendors implement standards. For an EDA vendor, supporting a standard is a business decision. In general, this means they support the features their customers request. Hence, you can influence the process by learning the new features and making the vendors aware of the ones that are important to you. The person with the most power is the person who funds your tool licenses. Make sure they are aware of what you need and make sure to forward your requests through them.

Index

Printed and bound by CPI Group (UK) Ltd, Croydon, CR0 4YY

03/10/2024

01040319-0002